T0198237

Jobsite First Aid: A Field Guide for the Construction Industry

Dan Johnson

Cengage

Australia • Brazil • Canada • Mexico • Singapore • United Kingdom • United States

Jobsite First Aid:
A Field Guide for the
Construction Industry
Dan Johnson

Vice President, Editorial:
Gregory L. Clayton

Director of Building Trades:
Taryn Zlatin McKenzie

Executive Editor:
Robert Person

Development Editor:
Nobina Preston

Marketing Director:
Beth A. Lutz

Marketing Manager:
Marissa Maiella

Marketing Coordinator:
Rachael Torres

Production Director:
Carolyn Miller

Senior Content Project
Manager: Stacey Lamodi

Art Director:
Benjamin Gleeksman

© 2012 Cengage Learning, Inc. ALL RIGHTS RESERVED.

No part of this work covered by the copyright herein may be reproduced or distributed in any form or by any means, except as permitted by U.S. copyright law, without the prior written permission of the copyright owner.

For product information and technology assistance, contact us at **Cengage Customer & Sales Support, 1-800-354-9706 or support.cengage.com.**

For permission to use material from this text or product, submit all requests online at **www.copyright.com**.

Library of Congress Control Number: 2010942283

ISBN: 978-1-111-03863-2

Cengage
200 Pier 4 Boulevard
Boston, MA 02210
USA

Cengage is a leading provider of customized learning solutions with employees residing in nearly 40 different countries and sales in more than 125 countries around the world. Find your local representative at: **www.cengage.com.**

Visit us at **www.InformationDestination.com**

To learn more about Cengage platforms and services, register or access your online learning solution, or purchase materials for your course, visit **www.cengage.com.**

Notice to the Reader
Publisher does not warrant or guarantee any of the products described herein or perform any independent analysis in connection with any of the product information contained herein. Publisher does not assume, and expressly disclaims, any obligation to obtain and include information other than that provided to it by the manufacturer. The reader is expressly warned to consider and adopt all safety precautions that might be indicated by the activities described herein and to avoid all potential hazards. By following the instructions contained herein, the reader willingly assumes all risks in connection with such instructions. The publisher makes no representations or warranties of any kind, including but not limited to, the warranties of fitness for particular purpose or merchantability, nor are any such representations implied with respect to the material set forth herein, and the publisher takes no responsibility with respect to such material. The publisher shall not be liable for any special, consequential, or exemplary damages resulting, in whole or part, from the readers' use of, or reliance upon, this material.

Printed in the United States of America
Print Number: 04   Print Year: 2022

TABLE OF CONTENTS

Every day, as numerous workers commute to their jobsites, any number of things may go through their minds—how to pay off bills, an argument they had with their spouse, the state of the economy, the health of their children, and the work that needs to be accomplished that day. Most likely they are not thinking about getting injured on the jobsite that day, but it is possible that some of them will. Nobody plans to get injured; that is why we call them accidents. But injuries do happen. According to the Bureau of Labor Statistics (www.bls.gov), over 4 million recordable non-fatal illnesses and injuries occur in the United States during the typical year. To break it down further, almost 11,000 non-fatal illnesses or injuries occur during a typical workday, 450 non-fatal illnesses or injuries take place during the typical lunch period, and 115 non-fatal illnesses or injuries transpire during a typical 15-minute morning or afternoon break.

These statistics show the risk associated with the workplace and the importance of maintaining workplace safety and health. Workplace safety and health is a very serious matter, but something that is taken for granted far too many times.

Think about if you are injured on the job:

© iStockphoto/aabejon

- Will someone on-site be prepared to help you?
- Will you receive the immediate care that may be required?
- What happens if you are injured while by yourself?

Now think about if an accident occurs to someone else on the jobsite:

- Will you know what to do?
- Will you respond?
- Who is counting on you to help?

These questions may be difficult to answer right now, but as you read through the pages of this guide the answers will become clearer.

Jobsite First Aid Guide: A Field Guide for the Construction Industry has been developed to prepare you in the event a jobsite injury or illness occurs. This guide will take you through a systematic approach to responding to emergency situations on your jobsite, whether that includes a life-threatening chemical exposure, a forklift rolling over, or a minor burn. You should note that this guide is not a substitute for hands-on training. Rather, it is designed to be kept in your tool box, glove box, or desk drawer as a reference to make sure that if you need to respond to an emergency, you are prepared.

As a volunteer first aid provider, your level of training will be lower than someone who is a professional rescuer. However, even though you are not paid to be a first aid provider, you still play a critical role—either in contacting Emergency Medical Services (EMS) or providing basic emergency medical care.

Image Copyright BryanJayne, 2012. Used under license from

CHAPTER 1

The Role of a First Aid Provider

Do you want to help people in need? If you see an accident or emergency situation develop, is your first impulse to see what you can do to help? If so, you may have the attitude and aptitude to be a first aid provider. However, you may have questions about what is involved with providing first aid in emergency situations. If so, the information in this guide should provide some answers.

WHAT IS A FIRST AID PROVIDER?

First aid is defined as emergency aid or treatment given to an injured or seriously ill individual before regular medical services arrive or can be reached. *Provider* is defined as a person who supports a family or another person. With these definitions, we can determine that a *first aid provider* is a person who supports another person by furnishing emergency aid or treatment before advanced medical services arrive.

First aid providers are not experts, nor are they expected to be experts. Their role is simply to help someone in need until advanced medical care arrives. First aid providers offer care within the guidelines of their training and to the best of their abilities, and then turn the patient over to the experts.

Used under license from shutterstock.com.

EMERGENCY MEDICAL SERVICES

As a first aid provider, you play a critical role in patient care. However, it is very important to understand that you are not alone. Rather, you are a vital part of the Emergency Medical Services (EMS) system. The EMS system, a network of people and organizations designed to provide aid in an emergency, includes doctors, nurses, paramedics, police officers, firefighters, and emergency operators, as well as many other support personnel. All members of this system are extremely proficient in their skill level and play a vital role during emergencies.

Individuals in need generally request the EMS system through an emergency number such as 911, which is used in most of North America. However, not all areas in the United States are covered by 911. Some remote areas use different emergency numbers in order to reach the EMS system, such as those for the local police department, fire department, hospital, or another emergency number. If you live in, work in, or travel to remote areas, make sure you know how to call for help. You should have the necessary emergency number written down in a prominent location and program that number into your mobile phone.

You should become familiar with your local EMS system. You may want to call them through their non-emergency number to learn how the system works. It is also worthwhile to ensure they know where your jobsite is located. As contractors build in new areas, it may take the EMS system a while to ensure their GPS mapping programs are fully updated. If an emergency takes place at your jobsite, you don't want to have precious minutes wasted while the paramedics are trying to find where you are located. Creating an open communication with the EMS system before starting a new job can become extremely valuable later on if there is an emergency at that location.

Did You Know?

Most countries do not use 911 as their emergency number. In fact, the European Union uses 112 as its emergency number, Australia uses 000, and Japan uses 110.

GOOD SAMARITAN LAWS

Many people choose to help others, whether they are strangers or not, simply out of kindness and concern. Unfortunately, we live in a very litigious society. It is hard to imagine patients deciding to sue their rescuer, but it has happened. Therefore, states have enacted various *Good Samaritan Laws* to protect first aid providers. In general, when first aid providers render aid in an emergency situation, the Good Samaritan Laws will protect them from any liability relating to their actions.

There are several important considerations regarding treatment as a first aid provider that you must keep in mind in order to be protected under Good Samaritan Laws:

- **Consent**. As a first aid provider, you must have the patient's consent to provide assistance. To gain consent, you should introduce yourself and then ask if it is okay for you to provide help. If the patient says yes, consent is granted. If the patient says no, you should not proceed. If you provide aid without consent, it can be considered assault.

Implied Consent

Consent may be implied if the patient is unconscious, delusional, or impaired in some way. If such a patient is suffering from a condition for which a reasonable person would give consent for treatment, consent is implied and the first aid provider can proceed.

- **Good faith**. You must act in good faith to help the patient. Acting in good faith means you act with honesty, decency, and fairness with no intent to defraud.
- **Grossly negligent**. You must act without gross negligence. Gross negligence, which means serious carelessness, is the opposite of acting in good faith.
- **Volunteer**. To be covered by Good Samaritan Laws, you must be a volunteer acting as a provider and must not be motivated to do so by material or financial gain.

Good Samaritan Laws are not applied universally within the United States. Rather, each state has its own standard. For this reason, it is a good idea to investigate the Good Samaritan Laws in your state for your own protection.

PERSONAL SAFETY

Your first priority as a first aid provider is to assure the scene is safe before entering to provide assistance. If you become a patient, the accident becomes much more complicated for everyone involved.

Image Copyright Karin Hildebrand Lau, © 2012. Used under license from Shutterstock.com

You should not enter an unstable accident area or otherwise unsafe scene before trained personnel have taken care of all the existing hazards such as spilled chemicals, poisonous gases, live electrical wires and parts, fires, and other hazards. If you find yourself already in a hazardous area, remove yourself and the patient from the area as quickly and safely as possible. If the patient is unable to move, leave the patient and get help. When entering an accident scene, you should look around to make sure conditions don't change, creating additional hazards. Always be on the lookout for unknown hazards (those hazards that aren't immediately recognized).

Since an emergency situation presents multiple hazards to you as a first aid provider, you should take steps to ensure your personal safety. You should also protect yourself from potential hazards such as blood and other bodily fluids when treating a patient. This subject will be discussed in more detail in Chapter 2.

YOU ARE NEEDED!

First aid providers are extremely important in our society. Not only are they relied upon to notify other members of the EMS system of emergencies, but having well-trained and responsive first aid providers available to supply basic care and treatment to an injured person in a critical situation may save a patient's life. As a first aid provider, regardless of the nature of the emergency, you are needed!

CHAPTER 2

Emergency Response/React

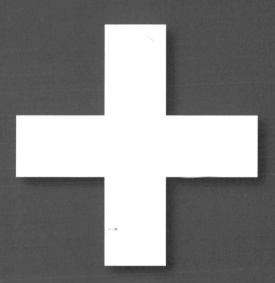

You are now continuing your journey toward becoming a first aid provider. When responding to emergencies it is very important to protect yourself from harm and the patient from further harm. There are numerous types of emergencies that may need your response, including but not limited to:

- Vehicular accidents
- Falls from heights
- Severe bleeding from a cut or other wound
- Puncture wounds
- Broken bones and sprains
- Back injuries
- Medical illnesses such as:
  - Heart attack
  - Stroke
  - Diabetes
  - Seizure

Your response in these emergencies will vary depending on the situation. As mentioned in Chapter 1, the first step to take in an emergency is providing the correct response. This chapter will take you through a step-by-step approach to emergency response. By following these steps, you will give the patient the best chance at survival. In an actual emergency, you may not have this guide with you. However, the steps we will go over are simple and easy to remember. All you need to do is **REACT.**

Step 1—**R**ecognize a medical emergency

Step 2—**E**valuate the environment/Think safety first!

Step 3—**A**ssess responsiveness

Step 4—**C**all for EMS

Step 5—**T**ake action

STEP 1. RECOGNIZE A MEDICAL EMERGENCY

The first step when responding to a medical emergency is recognizing that there is an emergency. Medical emergencies come in many different varieties. Some are obvious, such as a worker falling off the roof

© iStockphoto/mladn61

of a building. Others may be more subtle, such as a worker who has a diminished mental capacity due to chemical exposure.

Recognizing when a situation is a medical emergency can be difficult when people deny that their symptoms are serious and insist they just want to be left alone. Regardless of the situation, it is important for you to determine if it is indeed an emergency.

Some key indicators that might indicate an emergency are:

- Heavy construction equipment in an odd position, such as on its side
- A car with a door left open and no one around
- Screams, yells, or other loud noises
- Someone waving you down, trying to get your attention
- A vehicle that has physical damage
- A broken fence or other landscaping that appears to have been run over or hit
- A person lying on the ground or other abnormal location
- A person making strange noises or wandering around aimlessly
- A group of people gathering around in a peculiar fashion

These key indicators, combined with your common sense, will assist you in recognizing a medical emergency. As a first aid provider, you should respond to all suspected emergencies. Even if other people are present, your help may still be needed.

Therefore, if you see a large gathering of people, don't assume someone is already providing aid. Approach the group and offer to help unless EMS is at the scene.

STEP 2. EVALUATE THE ENVIRONMENT/ THINK SAFETY FIRST!

Once you have recognized that there may be an emergency, the next step is evaluating the environment. As you evaluate the environment, always think safety first!

To properly evaluate the environment, you will need to approach the scene to gather additional information. When approaching the scene, be aware of what is occurring in the environment. Never put yourself in a situation that might lead to injury or otherwise prevent you from providing first aid. Your top priority should be personal safety, since emergency scenes can be dangerous. Approach the scene cautiously. Do not run, since that may prevent you from recognizing hazards in the environment. If the scene is not safe, do not enter. Instead, call EMS for assistance.

There are several major hazards to be aware of when approaching a scene.

Traffic

Many first aid providers have been injured—or even killed—when responding to accidents. One major issue when responding to an accident near a roadway is oncoming traffic. Drivers traveling at high speed may not notice you providing care, especially if they are distracted by the accident scene. Take steps to ensure you are seen, and arrange for traffic to be rerouted or otherwise controlled. You should use extreme care when responding to a medical emergency near a roadway.

Fire or Smoke

Environmental barriers such as smoke or fire can make emergency response more difficult if not impossible. Not only do they obstruct the rescuer's view, but smoke and other pollutants can cause asphyxiation, whereas fire may lead to severe burns. Even firefighters who receive a great amount of ongoing training can be injured because of fire and/or smoke. Be very aware of these hazards and do not put yourself at risk.

Water

Never attempt a water rescue unless you have specific training and experience. Rushing water is an extremely powerful force that can sweep away vehicles and other large objects, especially during or after a major storm. Rivers and other bodies of water also have strong currents and other hazards that can complicate a rescue. *Never attempt a water rescue unless you have specific training and experience.*

Chemical

Chemical hazards—including liquids, gases, vapors, and fumes—can be hard to recognize. When approaching a scene where someone has collapsed, consider the possibility of a chemical exposure or lack of oxygen. If you suspect that chemicals caused the emergency or were released because of an accident, stay clear until you can acquire the appropriate protective equipment. In addition, only respond if you have been trained to respond to chemical emergencies.

Electrical

Be aware of any "hot" electrical devices (those that can provide a shock or supply current) that may have been damaged due to an accident or weather event, including downed power lines. You should assume all electrical equipment is live. Therefore, ensure qualified personnel de-energize electrical equipment before you provide first aid care.

Weather or Natural Disaster

You must use extreme caution when attempting to provide first aid during severe weather conditions or natural disasters. Both the environment and the hazards within it will constantly change. Be diligent in monitoring the situation and make adjustments to protect yourself as warranted.

Violence

Violence can occur anywhere, including an emergency scene. Others at the scene may be acting violently, or the patient may be out of control. The violence can be directed toward anyone—patients, bystanders, or even first aid providers. Be attentive to your surroundings and do not respond if there is ongoing violence.

Unknown

Consider the possibility of undetected dangers as you respond and be ready to alter your response if they arise.

Blood and Other Bodily Fluids

Some hazards develop from coming in contact with blood and other bodily fluids. The main hazard is disease transmission. Although the chances of getting a disease while providing first aid is minimal, you may be exposed to blood and other bodily fluids that are potentially infectious, so a risk does exist. The level of risk will vary depending on the virus or disease and the route of exposure. For this reason, you should take precautions to prevent contact with blood and other bodily fluids.

As a first aid provider, you should assume that all blood and bodily fluids at the emergency scene are infectious. The Centers for Disease Control and Prevention (CDC) recommends the use of Universal Precautions—protective barriers such as gloves, gowns, aprons, masks, or protective eyewear—which can reduce your risk of exposure to potentially infectious materials, especially on the skin and mucous membranes. Many first aid kits provide protective devices for Universal Precautions.

The two most common types of protection used during first aid are gloves and masks or shields.

Gloves

Disposable non-latex gloves will help you prevent exposure to blood and other bodily fluids as you provide first aid. You should wear gloves even when there is no obvious evidence of blood and other bodily fluids. All first aid kits should have at least two pairs of gloves. Inspect each glove prior to putting it on. If you notice any damage to the glove, replace it immediately. Ideally you should use non-latex gloves, such as those made from vinyl or nitrile, since you or the patients may be allergic to latex. Remember these important steps when using gloves:

- Use only undamaged gloves; replace them if necessary
- Replace gloves when working with different patients or when the gloves become heavily soiled

- Remove gloves carefully to prevent splattering blood and other bodily fluids
- Dispose of soiled gloves properly to help prevent exposure to potentially infectious material

Glove Disposal Tip

If you do not have an appropriate bodily fluid disposal system, ask the EMTs or paramedics if they will dispose of the gloves in their ambulance.

CPR Masks and Shields

When providing rescue breathing, a mask (Figure 2-1) or shield (Figure 2-2) will prevent you from making direct contact with the patient's mouth, reducing the possibility of coming in contact with blood and other bodily fluids. These masks and shields are for single use and should be disposed of properly after use.

Courtesy of iStockPhoto.com

FIGURE 2-1 CPR Mask with One-Way Valve

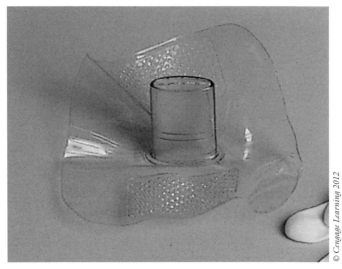

© Cengage Learning 2012

FIGURE 2-2 CPR Shield

Did You Know?

When you do not have immediate access to personal protective devices, you may want to use improvised barriers such as a plastic bag or clothing. You can find more information on Universal Precautions and bloodborne pathogens by visiting: **http://www.cdc.gov** or **http://www.osha.gov**.

STEP 3. ASSESS RESPONSIVENESS

Once you have determined that the scene is safe for you to enter, you must assess the patient's responsiveness. This is a simple process. First, you should determine the severity of the emergency. Keep alert for changes in the situation as you approach the patient. Once you have reached the patient, follow these steps.

Introduce Yourself

Tell the patient your name and level of training and ask if you can help. Say something like: "Hello, my name is Dan; I am trained in first aid. Can I help you?" If the patient responds to you in a meaningful manner, consider that a sign that the patient is responsive.

Table 2-1

Patient Response	First Aid Provider Action
Meaningful	Act to determine the severity of the patient's injury or illness by asking questions (covered in future chapters). If the patient has a serious illness or major impact injury, call EMS.
Not meaningful	Go to the next method (Tap and shout).

Tap and Shout

If the patient does not respond to your introduction, tap the patient on the shoulder, shout your name, and ask if you can provide help (Figure 2-3). Your intent should be to startle the patient and illicit a response.

Table 2-2

Patient Response	First Aid Provider Action
Meaningful	Act to determine the severity of the patient's injury or illness by tapping the patient on the shoulder and asking if help is needed. If the patient has a serious illness or major impact injury, call EMS.
Not meaningful	Go to the next step (Gain consent and call EMS).

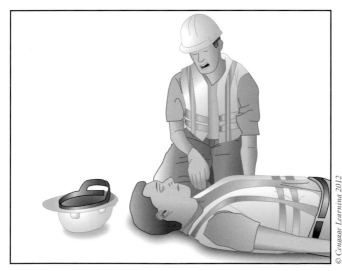

FIGURE 2-3 Tap and shout

Gaining Consent

As covered in Chapter 1, you need to gain consent for care from a responsive patient. When you introduce yourself and ask the patient if you can help, you are asking for consent. If the patient responds by saying "Yes," you have been given consent. If the patient responds by saying "No," consent has been denied and you should stop without providing any help. Even if no consent has been granted, it is a good idea to move on to the next step and call EMS. If you are dealing with an unresponsive patient, consent is generally implied, making it acceptable to proceed.

STEP 4. CALL FOR EMERGENCY MEDICAL SERVICES

The fourth step is to call for Emergency Medical Services (EMS), which was explained in Chapter 1. As the first aid provider, you must ensure advanced medical help responds to the scene as soon as possible. Once you have assessed the patient's responsiveness, you will call EMS for the following reasons:

- **Unresponsive patient.** If the patient does not respond to your tap and shout, call EMS immediately.
- **Major impact injury.** If you determine that a responsive patient has suffered a major impact injury (as presented in Chapter 8), call EMS at once.
- **Serious illness.** If you determine that a responsive patient has an emergency sign of illness (as presented in Chapter 9), call EMS without further delay.

If others are at the emergency scene, ask someone (identify a particular person) to call EMS. This will free you up to provide basic care to the patient while EMS is being called. If you have someone call EMS for you, make sure that the person actually calls EMS. If the designated caller has a cell phone, instruct the person to call EMS in front of you. If the caller needs to leave the scene to call EMS, ask the person to return to you once the call is made.

If you are alone with the patient, use your cell phone to call EMS. If you do not have a cell phone, you will need to leave the patient and call EMS. *Do not delay calling EMS.*

Always Remember

Do not delay in calling EMS once you have determined that EMS is needed, since their response times vary throughout the country. The sooner advanced medical help is on their way, the better chance the patient has for survival.

STEP 5. TAKE ACTION

You have now responded to the emergency, evaluated the environment for safety, accessed the patient's responsiveness, and called for EMS. The final step is to take action. As a first aid provider, your role is vital. In the time it takes EMS to drive to the scene you may need to provide basic care to the patient after a major impact injury or help the patient remain calm and comfortable during a serious illness. Your actions may include providing CPR, choking management, control of bleeding, or first aid for shock. In the next few chapters you will learn about these and other situations where your help may be needed. By taking action you may help save a life!

The entire REACT process will probably take you less than 60 seconds to complete; however, following these basic steps will ensure your safety and give the patient a better chance of survival.

© iStockphoto/Abejon Photography

CHAPTER 3

ABCs of Life Support

The human body requires certain functions to remain ongoing in order to sustain life. An unresponsive patient may have underlying issues causing the illness or injury; however, as the first aid provider, you should not concern yourself with the underlying cause. Your focus should be on the ABCs of life support: airway, breathing, and circulation.

AIRWAY

The first part of the ABCs of life support is the airway. If a patient has an open airway, the opportunity exists for the patient to breathe. As a first aid provider, you may find unresponsive patients lying on their back or in some other position, such as on their stomach or side.

Patients Not on Their Backs

If you respond to a patient who is not lying on their back, the patient's airway may be open. Assess the airway by determining if the patient is breathing. Look for signs of obvious breathing. If an unresponsive patient is breathing in an effective manner, you can assume that the airway is open and there is no need to move the patient at this time. However, continually monitor the airway to ensure that it stays open. If the patient is not obviously breathing, you will need to roll the patient over carefully, onto the patient's back, to manually open the airway with the head-tilt, chin-lift maneuver (this technique is described later in this chapter).

Patients on Their Backs

If unresponsive patients are found on their backs, they will commonly have blocked airways. This occurs from the tongue falling against the back of the throat. You will need to manually open the airway, which can be done by using the head-tilt, chin-lift maneuver.

Head-Tilt, Chin-Lift Maneuver

To manually open the airway of an unresponsive patient using the head-tilt, chin-lift maneuver, follow these steps:

- Make sure the patient is lying against a firm surface, with the patient's back flat against the surface.
- Position the palm of one of your hands on the patient's forehead and place the fingers of your other hand under the patient's chin (Figure 3–1).
- Using the palm of your hand, push the patient's head back while gently lifting up on the patient's chin (Figure 3–2).

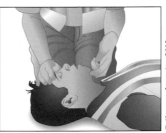

© Cengage Learning 2012

FIGURE 3–1

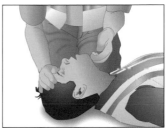

© Cengage Learning 2012

FIGURE 3–2

The head-tilt, chin-lift maneuver is designed to pull the tongue away from the patient's throat and open the airway. When using this technique, avoid putting pressure on the soft tissue beneath the patient's chin and keep the mouth somewhat open.

Once you have opened the airway, you will need to assess the patient's breathing. However, prior to that, you may run into a couple of obstacles.

Clearing Solids from the Airway

If there is a solid object found in the patient's airway, remove it with a gloved finger by doing a finger sweep. Make sure you do not push on the object, as that may thrust the object further down into the airway (Figure 3–3). If the object is too far down into the airway, proceed with CPR as shown in Chapter 4.

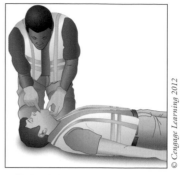

FIGURE 3–3

Clearing Fluids from the Airway

Fluids can block the patient's airway and prevent breathing. You should act quickly if you find fluids in the mouth of an unresponsive patient. A very effective and practical method, called the log roll, uses the following steps to clear fluids from a patient's mouth:

- Make sure the patient is lying against a firm surface, with the patient's back flat against the surface.
- Kneel down next to the patient.
- Raise the patient's arm closest to you above the patient's head (Figure 3–4).

FIGURE 3–4

- Place your hand under the patient's neck and head.
- Place your other hand on the elbow of the patient's arm farthest from you.

- Roll the patient toward you as a unit, making sure to support the patient's neck and head (Figure 3–5).
- Place the patient in a side-lying position, as that will help the fluid drain from the patient's mouth.
- Use your gloved finger to remove any other fluids from the patient's mouth by doing a finger sweep (Figure 3–6).
- Return the patient to his or her back and use the head-tilt, chin-lift maneuver to open the airway.

Maintaining an Open Airway

Once you have achieved an open airway in the patient, make sure you maintain it using the head-tilt, chin-lift maneuver (Figure 3–7). If you release your hands, the patient's tongue could block the airway again.

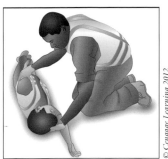

FIGURE 3–5

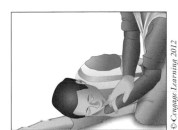

FIGURE 3–6

FIGURE 3–7

© Cengage Learning 2012

Recovery Position

Uninjured patients who are breathing on their own may be placed in the recovery position to maintain an open airway. By placing an uninjured and breathing patient in the recovery position, you will be able to provide other critical care, such as bleeding control and first aid for shock (as shown in Chapters 6 and 7, respectively). To place an uninjured and breathing patient in the recovery position, follow these steps:

FIGURE 3–8

- Place the patient against a firm surface on the ground, with the patient's back flat against the surface.
- Kneel down next to the patient.
- Raise the patient's arm closest to you above the patient's head.
- Take the patient's arm that is farthest from you and bring it across the patient's chest, placing the patient's hand on top of the shoulder (Figure 3–8).

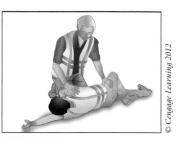

FIGURE 3–9

- Grasp the patient's body with both hands and carefully roll the patient as a unit toward you. The patient will rest on the ground with the knee and elbow touching the ground (Figure 3–9).
- Make sure to roll the patient forward enough to maintain an open airway.
- Monitor the patient to make sure breathing continues (Figure 3–10).

FIGURE 3–10

© Cengage Learning 2012

BREATHING

The next part of the ABCs of life support is breathing. Once you have opened the patient's airway, it is time to assess the patient for breathing. To assess a patient's breathing, kneel down next to the patient and look, listen, and feel for breathing (Figure 3–11).

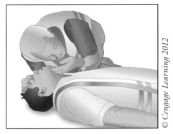

© Cengage Learning 2012

FIGURE 3–11

Look for the patient's chest to rise with each breath.

Listen for breathing by getting your ear very close to the patient's mouth.

Feel for exhaled breath against your cheek or ear.

When you are assessing for breathing, you should be assessing for normal breathing. This means that the patient is breathing the way a person normally breathes—steady inhalations, followed by solid exhalations. If the patient is gasping, infrequently sighing, or taking other short ineffective breaths, the patient should not be considered to be breathing.

No Breath

If you do not see signs of breath or you determine the patient has ineffective breathing, you should *immediately perform CPR* as shown in Chapter 4. Do not move on to assessing the patient's circulation. If the patient is not breathing, the best care you can provide is CPR.

Breathing Is Present

If you determine the patient has normal breathing, you should maintain the patient's airway using the head-tilt, chin-lift maneuver and then move on to assessing for circulation.

CIRCULATION

Assessing the patient's circulation is the final step of the ABCs of life support. By this point you have called for EMS, opened the patient's airway, assessed for breathing, and found the patient to be effectively breathing. While maintaining the patient's airway using the head-tilt, chin-lift maneuver, you should begin to assess for circulation.

Since you have already determined that the patient is breathing normally which is a sign of circulation, there is no need to check for a pulse. You should assess for circulation through the following steps:

- **Examine the rest of the patient's body for signs of severe bleeding.** If severe bleeding is found, make sure the airway remains open before starting to control the bleeding.
- **Assess the patient's tissue color and skin temperature to look for signs of shock.** if you find signs of shock, you should make sure the airway remains open before starting first aid for shock.

CONTINUAL ASSESSMENT

The patient may take a turn for the worse at any time. Because of this, you should continually assess the ABCs of life support: airway, breathing, and circulation.

CHAPTER 4

CPR for Cardiac Arrest

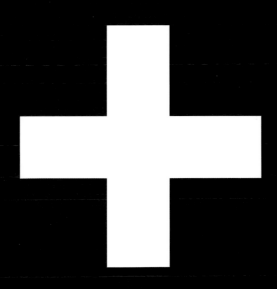

Sudden cardiac arrest, the termination of normal blood circulation due to the heart's failure to function effectively, is a major medical emergency you may need to respond to as a first aid provider. Sudden cardiac arrest is generally caused by an abnormality in the patient's heart rhythm due to a breakdown in the heart's electrical system.

When normal blood circulation is terminated, oxygen will not be effectively delivered to the body. If less oxygen is delivered to the brain than is needed, the result can be the loss of consciousness, which then causes irregular or absent breathing. Permanent brain damage is likely if sudden cardiac arrest goes untreated.

In certain situations, sudden cardiac arrest is reversible if treated early. The treatment for sudden cardiac arrest is cardiopulmonary resuscitation (CPR), followed by defibrillation. Since automated external defibrillators (AED) are now commonly found in many public places, including airports, shopping malls, and sports and entertainment venues (Figure 4-1), many more sudden cardiac arrest patients can have defibrillation treatment much sooner after they collapse.

FIGURE 4–1 AED signs appear in public venues

Courtesy of iStockphoto.com

CPR and AED training is recommended for all first aid providers. Formal training can assist you in becoming more comfortable with the elements of CPR and usage of an AED. However, even individuals that have had no formal training in these areas can greatly assist a victim of sudden cardiac arrest.

SEQUENCE FOR SURVIVAL

The Sequence for Survival (Figure 4-2) indicates the essential components that provide a victim of sudden cardiac arrest with the greatest chance of survival:

© Cengage Learning 2012

FIGURE 4–2

- **Early EMS.** Remember "REACT" and call EMS immediately.
- **Early CPR.** Perform CPR as soon as EMS has been called.
- **Early defibrillation.** If available, get an AED and use it without delay.
- **Early advanced care.** EMS will provide advanced medical care when they arrive on the scene.

A weak component in the Sequence for Survival greatly reduces a victim's chance for surviving sudden cardiac arrest. As a first aid provider, your role is to call EMS, provide CPR, and use an AED. All of these components can be done without formal training.

CPR

It is estimated that less than 50% of patients suffering from sudden cardiac arrest receive CPR. Reasons why may include lack of training, apprehension by people because they are afraid they will hurt the victim, and not having the appropriate CPR mask. In 2010, the American Heart Association produced new guidelines aimed at overcoming these concerns and increasing the percentage of sudden cardiac arrest patients who receive CPR.

CPR consists of three components: compressions, airway, and breathing (CAB). Always remember the steps of "REACT" when responding to any emergency.

Compressions

Compressions refer to pushing on the patient's chest to squeeze the heart between the breast bone and back. This pushes blood throughout the body to sustain life. Compressions have been found to be the most important step in CPR. To conduct compressions on an adult, follow these steps:

- Make sure the patient is placed on a flat, firm surface, with the patient's back flat against the surface.
- Kneel down next to the patient's chest.
- Place the heel of your first hand directly over the breast bone in the middle of the patient's chest right between

the nipples and your other hand on top of the first hand (Figure 4-3A). Interlock your fingers to maintain your hand position on the chest (Figure 4-3B).

- Place your knees as close to the patient as possible and bring your shoulders directly over your hands, making sure to lock your elbows (Figure 4-3C).

- Using the weight of your body, push hard and fast to complete a compression. You should push down at least 2 inches (Figure 4-3D). Allow the patient's chest to fully expand back to its normal position, keeping your hands in constant contact with the patient's breast bone.

- Compress the patient's chest at a smooth and constant rate of at least 100 times per minute. The rate of 100 times per minute is similar to the beat of the Bee Gees' song, "Staying Alive." It may be beneficial to sing this song in your head to keep the appropriate rate. Always remember to push hard and fast while giving compressions.

You should feel and hear snaps and pops while performing compressions. These are natural, common noises that should not distract you from giving compressions.

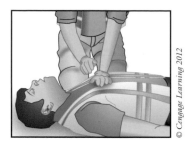

FIGURE 4–3A

FIGURE 4–3B

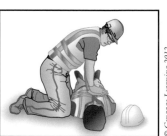

FIGURE 4–3C

FIGURE 4–3D

© Cengage Learning 2012

Airway

Use the head-tilt, chin-lift maneuver (Figure 4-4) to open the patient's airway, as presented in Chapter 3.

Breathing

Breathing directly into a patient's mouth using mouth-to-mouth resuscitation or breathing through a CPR mask provides oxygen to the patient. This oxygen will be conveyed into the bloodstream, providing oxygenated blood to sustain the patient's life. Follow these steps to breathe for a patient:

- Use a CPR mask or CPR shield whenever possible to prevent transmission of disease.
- Use the head-tilt, chin-lift maneuver to open the airway.
- If using a CPR mask:
  - Place the mask over the patient's mouth, covering the mouth and nose (Figure 4-5).
  - Press the mask against the patient's mouth to create a seal while maintaining the open airway (Figure 4-6).
- If using a CPR shield:
  - Place the shield over the patient's mouth, inserting the valve into the mouth (Figure 4-7).
  - Pinch the patient's nose, maintain the open airway, and

FIGURE 4–4

© Cengage Learning 2012

FIGURE 4–5

© Cengage Learning 2012

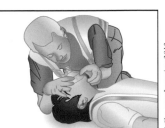

FIGURE 4–6

© Cengage Learning 2012

FIGURE 4–7

© Cengage Learning 2012

use your lips to create a seal over the patient's lips (Figure 4-8).

- Mouth-to-mouth resuscitation can be performed in the same manner as when using a CPR shield—just do not use the shield.
- Take a normal breath and blow into the patient's mouth, watching for the patient's chest to rise as you breathe (Figure 4-9). The breath should be about 1 second in length.
- Break the seal and allow for exhalation before giving another breath (Figure 4-10).

FIGURE 4–8

FIGURE 4–9

AUTOMATED EXTERNAL DEFIBRILLATOR

An automated external defibrillator (AED) is a portable electronic device that automatically diagnoses dangerous heart rhythms and delivers a shock to the patient to return the heart to an effective rhythm. This very useful medical device can help victims of sudden cardiac arrest who are experiencing certain heart arrhythmias or ventricular fibrillation.

FIGURE 4–10

Strong evidence exists that the use of AEDs by first aid providers has significantly increased the survival rate of sudden cardiac arrest victims. Like in other situations, it is not important for you, as a first aid provider, to diagnose whether or not the patient is experiencing ventricular fibrillation. It is, however, important for you to take action. An AED can be operated with minimal knowledge and no formal training.

If you respond to a medical emergency, find an unresponsive patient, and have an AED available, you should use the AED as part of CPR without delay. AED usage is a vital component of the Sequence for Survival. Always remember the steps of "REACT." AEDs have many

© Cengage Learning 2012

safety functions built in and will not allow an electrical shock to be delivered to a patient who does not have the appropriate heart arrhythmia.

To use an AED, follow these steps:

- Make sure the patient is lying flat against a firm surface, with the patient's back flat against the surface. You should continue CPR until the AED is ready to be used.
- Open the case of the AED and turn on the power to the AED (Figure 4-11).
- Listen for and follow the voice prompts of the AED, which will guide you through the proper operation. Some AEDs will have the defibrillator pads already connected, whereas other AEDs require you to open the defibrillator pads and plug the pads into the AED unit.

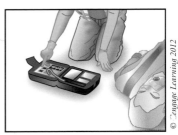

FIGURE 4–11

- Remove all clothing covering the patient's chest. If needed, cut the clothing to bare the chest (Figure 4-12).
- Dry the patient's chest if wet or sweaty, as defibrillator pads may not adhere well enough to deliver a shock when the patient's chest is wet. If excessive chest hair is present, use a razor to quickly shave the hair so the pads can properly adhere to the patient's chest.

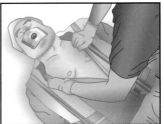

FIGURE 4–12

- Place the defibrillator pads on the patient's bare chest. The pads will have pictures on them showing you how and where to place them. Make sure the pads adhere to the skin by pressing hard on the pads, especially around the edges.
  - One pad will be placed below the patient's right collarbone, to the right of the breast bone and above the nipple (Figure 4-13).

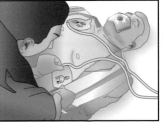

FIGURE 4–13

© Cengage Learning 2012

- The other pad will be placed on the patient's left side, below the nipple and over the ribs. The pad will be a few inches below the patient's armpit.
- The AED will automatically begin to analyze the patient's heart rhythm once the pads are properly connected and plugged in. Do not touch the patient while the AED is analyzing the patient's heart rhythm.
- If the AED determines that a shock is needed, it will alert you that "Shock is advised." Make sure no one is touching the patient—tell everyone to stay clear. When the AED is ready, push the shock button (Figure 4-14).

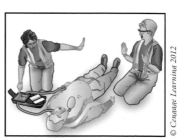

FIGURE 4–14

- Once a shock is delivered and it is safe to touch the patient, resume CPR. Continue CPR until the AED directs you to stop, the patient begins to move independently, or you are directed to stop by EMS personnel (Figure 4-15).
- If the AED determines that additional shocks are needed, deliver additional shocks as directed.

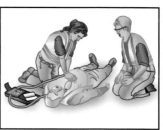

FIGURE 4–15

AEDs are extremely user friendly. If you have any problems during operation the AED, through voice prompts, will alert you and provide instructions to overcome those problems.

AEDs usually come equipped with all the supplies you need to use them. These supplies include gauze pads to dry the patient's chest, a razor to shave the chest, gloves, a CPR mask, extra defibrillator pads, an extra battery, and other necessary supplies. Organizations that have AEDs available for use need to ensure that they follow the manufacturer-prescribed maintenance program. AEDs should be ready for use at all times for maximum benefit.

PUTTING IT ALL TOGETHER

In this chapter, you have learned about CPR and AEDs, including how to perform each technique. Now you will learn the proper steps to put it all together. When a victim suffers sudden cardiac arrest, you play a critical role in sustaining the patient's life while waiting for EMS to respond.

As discussed earlier, you can perform CPR whether or not you are formally trained. However, there are some differences in the care you can provide based on your level of training. If you have no formal training, you should perform hands only CPR. If you have formal CPR training, you can choose to conduct full CPR or hands only CPR. The processes shown in this guide are for adults only.

Hands Only CPR

Hands only CPR is designed to reduce your apprehension as a first aid provider and promote the opportunity for more non-trained individuals to provide CPR. With hands only CPR, you will not perform any breathing for the patient. If you have no formal training or you are not comfortable giving breaths, follow these steps to properly perform hands only CPR:

- Perform the steps of "REACT".
- Call EMS immediately:
  - If you are by yourself, you should call EMS before providing further care. It is extremely important to get EMS on their way to the scene.
  - If you are not by yourself, send someone to call EMS, and then begin further care—make sure the person you send to call EMS actually calls.
- Retrieve the AED (if there is an AED available):
  - If you are by yourself, you should leave the patient to retrieve the AED.
  - If you are not by yourself, send someone to retrieve the AED, and then begin further care.
- Start compressions (Figure 4-16):
  - Start providing compressions at a rate of at least 100 per minute and compressing the patient's chest at least 2 inches.

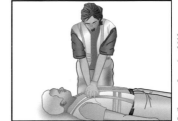

FIGURE 4–16

© Cengage Learning 2012

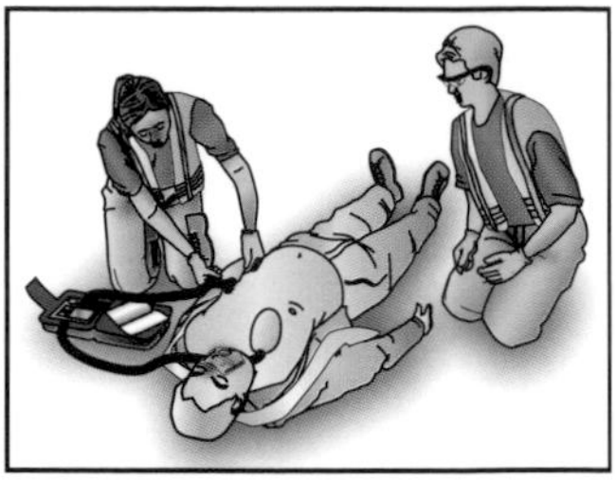

- Discontinue compressions only if an AED is ready to use, the patient begins to move independently, or you are directed by EMS personnel.
- Use the AED (Figure 4-17):
  - Use an AED as soon as it is available and follow all its voice prompts and instructions.

FIGURE 4–17

- Continue compressions:
  - Continue to provide compressions at a rate of at least 100 per minute and compressing the patient's chest at least 2 inches.
  - Discontinue compressions only if the AED directs you to stop, the patient begins to move independently, or you are directed to stop by EMS personnel.

Full CPR

Full CPR is recommended to any first aid provider who has had formal CPR training. The 2010 CPR guidelines have placed more emphasis on compressions while performing CPR. The new guidelines prioritize CPR into a new compressions–airway–breathing (CAB) format. You should begin with compressions prior to moving on to airway and then breathing. To perform full CPR, follow these steps:

- Perform the steps of "REACT".
- Call EMS immediately:
  - If you are by yourself, you should call EMS before providing further care. It is extremely important to get EMS on their way to the scene.
  - If you are not by yourself, send someone to call EMS, and then begin further care—make sure the person you send to call EMS actually calls.
- Retrieve the AED (if there is an AED available):
  - If you are by yourself, you should leave the patient to retrieve the AED.

- If you are not by yourself, send someone to retrieve the AED, and then begin further care.
- Start compressions (Figure 4-18)
  - Start providing compressions at a rate of at least 100 per minute and compressing the patient's chest at least 2 inches.
  - Administer 30 compressions.
- Open the patient's airway (Figure 4-19):
  - Use the head-tilt, chin-lift maneuver to open the patient's airway.
- Give breaths (Figure 4-20):
  - Give the patient two breaths lasting about 1 second each.
  - Monitor the patient's chest to ensure the chest rises while giving breaths.
  - Breaths should be given swiftly to minimize time between compressions.
- Continue the CAB format of CPR (Figure 4-21):
  - Provide 30 compressions.
  - Open the patient's airway.
  - Provide two breaths.
  - Discontinue CPR only if an AED is ready to use, the patient begins to move independently, or you are directed to stop by EMS personnel.

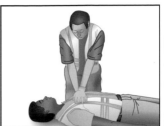

FIGURE 4–18

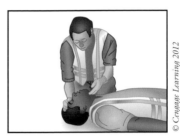

FIGURE 4–19

FIGURE 4–20

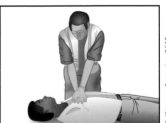

FIGURE 4–21

© Cengage Learning 2012

- Use the AED (Figure 4-22):
  - Use an AED as soon as it is available and follow all voice prompts and instructions.
- Continue CPR (Figure 4-23):
  - Continue to provide CPR with 30 compressions and two breaths.
  - Discontinue CPR only if the AED directs you to stop, the patient begins to move independently, or you are directed to stop by EMS personnel.

FIGURE 4–22

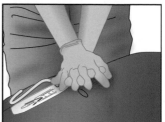

FIGURE 4–23

© Cengage Learning 2012

CHAPTER 5

Choking Management

Choking is a potentially life-threatening situation that must be treated seriously. Choking is usually caused by the blockage of the patient's airway by some foreign object. Although many foreign objects can cause choking, most often the cause is inadequately chewed food. The blockage in the airway cuts off the supply of oxygen which leads to trouble in breathing, which will eventually stop altogether if the blockage is not removed. As discussed in Chapter 4, lack of oxygen can cause brain damage and ultimately death. Therefore, it is imperative to act quickly when responding to a patient who is choking.

Remember the steps of "REACT" while responding to a patient who is choking. The universal sign for choking is when a person has the hands clutched over the throat.

Step 1—**R**ecognize a medical emergency

Step 2—**E**valuate the environment/Think safety first!

Step 3—**A**ssess responsiveness

Step 4—**C**all for EMS

Step 5—**T**ake action

CHOKING MANAGEMENT TECHNIQUES FOR A RESPONSIVE PATIENT

- Ask if the patient is choking.
  - If the patient can speak, cough forcefully, or breathe independently, stay nearby while the patient tries to clear the foreign object from the throat. The choking victim may be embarrassed from choking in front of others and might try to leave the room. However, do not allow the patient to leave the room.
- If the patient has difficulty breathing, speaking, or coughing, call EMS without delay.
- Ask for consent to help the patient: "Can I help?"
- Once you are given consent, position yourself behind the patient and perform abdominal thrusts.

Abdominal Thrust Technique

- Wrap your arms around the patient's waist (Figure 5-1).
- Make a fist with one hand and place it thumb-side against the abdomen along the midline of the patient and slightly above the belly button.
- Grasp the fist with the other hand (Figure 5-2).

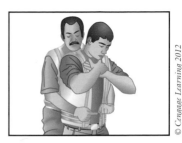

FIGURE 5-1

- Give five quick upward thrusts as if you are lifting the patient up. Each abdominal thrust should be given with the intention of clearing the obstruction.
- If done successfully, the obstruction will pop out of the patient's throat or become dislodged enough for the patient to extract it.

FIGURE 5-2

- If the object is not dislodged, continue performing abdominal thrusts until the obstruction is cleared or the patient becomes unresponsive.

CHOKING MANAGEMENT TECHNIQUES FOR AN UNRESPONSIVE PATIENT

If the patient becomes unresponsive while you are performing abdominal thrusts or the patient is found unresponsive; follow these steps:

- Follow the steps of "REACT."
- Make sure EMS has been called.
- Follow the ABCs of life support (airway, breathing, and circulation) and start CPR.
- Every time you open the airway, inspect the patient's mouth for a foreign body that may have become dislodged from the patient's throat.

© Cengage Learning 2012

- If a foreign object is found, use your gloved finger to sweep the object out of the patient's mouth—be careful not to push the object further into the throat (Figure 5-3).
- Continue CPR until the patient shows signs of circulation or until EMS arrives and takes over care of the patient.

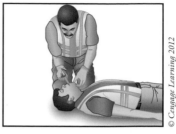

FIGURE 5-3

FOLLOW-UP MEDICAL CARE

Whenever you perform abdominal thrusts on a patient, make sure that the patient seeks out medical care. Even properly performed abdominal thrusts may cause internal damage. If not treated for the potential internal damage, the patient may experience potentially life-threatening complications.

CHAPTER **6**

Control of Bleeding

Bleeding commonly occurs as a result of jobsite accidents. Therefore, as a first aid provider, you should be ready to provide care to a bleeding patient. The body has a natural blood clotting process that will slow and eventually stop the flow of blood. Minor bleeding will often cease by itself in about 10 minutes. Major bleeding, however, can become a life-threatening condition requiring urgent attention if it does not stop. When treating a patient with major bleeding, you will need to provide care to help the blood clotting process (Figure 6-1).

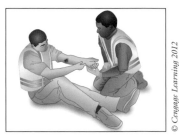

FIGURE 6-1

© Cengage Learning 2012

As discussed in Chapter 2, blood and other bodily fluids are potentially dangerous to you as a first aid provider. It is important to follow Universal Precautions as recommended by the Centers for Disease Control and Prevention (CDC) when treating these patients. Further information on Universal Precautions can be found at the CDC's website: http://www.cdc.gov.

Bleeding may be either external (obvious bleeding that breaks the skin's surface) or internal (bleeding within the body that cannot be seen by the naked eye). This chapter will discuss external bleeding and its control, whereas Chapter 7 will discuss the potentially life-threatening impact of internal bleeding.

TYPES OF BLEEDING

As a first aid provider, you should mainly be concerned about three different types of bleeding:

- **Arterial bleeding**—This severe and hard-to-control bleeding comes from an artery. An artery is a major blood vessel that carries oxygen-rich blood from the heart and distributes that blood throughout the body. Bright red blood that spurts with each beat of the heart is evidence of arterial bleeding. This type of bleeding is life-threatening and requires immediate attention.
- **Venous bleeding**—This bleeding comes from a vein. Veins carry blood back to the heart after the oxygen is delivered to the body. Venous blood is characterized by a stable flow and a dark red or maroon color. Venous bleeding is easier to control

than arterial bleeding; however, it can still be a life-threatening condition and should be treated as such.

- **Capillary bleeding**—This bleeding comes from the smallest of our body's blood vessels. Blood flow from a capillary is usually slow and oozes from the wound.

CONTROL OF BLEEDING

When you respond to a scene where a patient is bleeding, you should first remember the ABCs of life support: airway, breathing, and circulation. If the patient has an open airway and is breathing, you can then focus on bleeding control. Do not let the sight of blood prevent you from proving the ABCs of life support. Your three goals for controlling bleeding are:

- Stop the flow of blood.
- Prevent infection.
- Prevent shock.

Follow these steps to control the bleeding:

- Follow the steps of "REACT."
  - Recognize a medical emergency
  - Evaluate the environment
  - Assess responsiveness
  - Call for EMS
  - Take action
- If there is arterial bleeding, always call EMS.
- Monitor the ABCs of life support.
- Protect yourself by wearing gloves and other protective equipment (Figure 6-2).

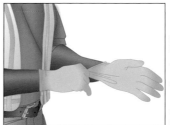

FIGURE 6-2

© Cengage Learning 2012

- Use a clean dressing and apply pressure directly over the wound (Figure 6-3).
  - During an emergency, the goal is to prevent the loss of blood. Blood will naturally coagulate. Using clean dressings with sterile gauze and direct pressure will speed up the coagulation.

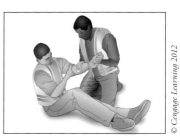

FIGURE 6-3

© Cengage Learning 2012

- If the blood soaks through the dressing, keep it in place and apply more dressings over the old dressing. The less a bleeding wound is bothered, the simpler it will be to control the bleeding (Figure 6-4). You may need to apply multiple levels of dressings to control the bleeding.

FIGURE 6-4

- Elevating the wound while applying direct pressure can also help you control the bleeding. Make sure you only elevate the wound if you do not suspect a spinal injury or fracture.

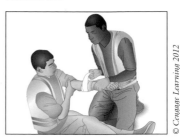

FIGURE 6-5

- Once the bleeding is controlled, apply a pressure bandage to the wound without removing any of the dressings (Figure 6-5).

   - A pressure bandage such as roll gauze is used to hold the dressings in place. While applying the bandage, make sure to use some force to keep pressure on the wound. After the pressure bandage is in place, check the area around the dressing to make sure that you did not cut off circulation. A slow pulse rate or bluish tissue around the wound or in the fingertips or toes may indicate that a pressure bandage is hindering circulation.

You should keep the ABCs of life support in mind while treating a patient for bleeding. Be ready for the circumstances to change. Loss of blood can cause people to lose consciousness, which may eventually require you to perform CPR.

BLOOD SPILLS

You must continue to protect yourself even after the patient's bleeding is controlled. There will often be blood spills that need to be cleaned up. Treat all blood spills as if they are infectious. If a blood spill does occur, follow these steps:

- Clean up the spill as soon as possible after it occurs. This will help to prevent others from coming in contact with the blood.

- Use disposable gloves at all times when cleaning up blood spills. If available, use a bloodborne pathogen kit or body fluid cleanup kit (Figure 6-6).
- Wipe up the blood with absorbent materials such as paper towels.

FIGURE 6-6

- Deluge the affected area with a solution of bleach and water consisting of ¼ cup of bleach to every 1 gallon of water.
- Let the solution settle on the affected area for approximately 30 minutes.
- Clean up the solution with absorbent materials.
- While your hands are still protected by gloves, place all of the contaminated materials in a red biohazard bag (Figure 6-7).

FIGURE 6-7

- Remove your gloves carefully to prevent splattering and place them in a biohazard bag.
- Dispose of the biohazard bag properly to prevent others from becoming exposed to the contaminated material.

REMOVING GLOVES

As previously mentioned, it is important to remove your gloves carefully to prevent them from splattering blood on yourself or others.

Remove the First Glove

- Pinch the glove just below the wrist, being very careful to only touch the outer surface of the glove.
- Pull the glove down in the direction of your fingers and begin turning the glove inside out.

© Cengage Learning 2012

- With your gloved hand, finish removing the first glove and hold that glove in the palm of your gloved hand (Figure 6-8).

Remove the Second Glove

- Put the index finger on your ungloved hand under the cuff of the second glove, being careful to touch only the inside portion of the glove.
- Pull down in the direction of your fingers and begin turning the glove inside out.
- Continue to pull the glove down and completely turn the second glove inside out. The first glove should be inside the second glove (Figure 6-9).

FIGURE 6-8

FIGURE 6-9

After Removing Your Gloves

- Dispose of the gloves in a red biohazard bag.
- Wash your hands thoroughly with antimicrobial soap and warm water for at least 2 minutes.

Control of bleeding is a very important task for a first aid provider. Remember to continue to monitor the patient for bleeding and be ready to act if circumstances change. Always keep in mind that bleeding can be life threatening.

CHAPTER 7

First Aid for Shock

Shock is a potentially life-threatening condition that occurs when a patient's organs are not receiving enough oxygenated blood. Shock may result from trauma, heatstroke, blood loss, an allergic reaction, severe infection, poisoning, severe burns, or other causes. If not recognized and treated properly, shock can result in organ damage and/or death. As a first aid provider, you should consider shock management a high priority. Although it is very difficult to diagnose shock away from the hospital, you should be able to recognize the various signs and symptoms related to shock.

SIGNS AND SYMPTOMS OF SHOCK

Although it is not imperative for you to diagnose why a patient may be going into shock, recognizing the following signs and symptoms and then providing appropriate treatment can greatly assist the patient in surviving shock:

- Cool and clammy skin that may appear gray or pale
- Weak and rapid pulse
- Shallow and slow breathing or rapid and deep breathing
- Low blood pressure
- Nausea or vomiting
- Lackluster eyes, in which the pupils may be dilated
- Feeling faint, weak, or confused
- Anxiousness or overexcitement
- Internal bleeding; the following are signs and symptoms of internal bleeding:
  - Bruised, swollen, tender, or rigid abdomen
  - Bruises on chest or signs of fractured ribs
  - Blood in vomit
  - Wounds that have penetrated the chest or abdomen
  - Abnormal pulse and difficulty breathing
  - Cool, moist skin

TREATMENT FOR SHOCK

As previously shown, there are many signs and symptoms for shock. If you suspect shock:

- Follow the steps of "REACT."
  - Recognize a medical emergency
  - Evaluate the environment
  - Assess responsiveness
  - Call for EMS
  - Take action
- Always call EMS.
- Monitor the ABCs of life support (airway, breathing, and circulation).
- Put the patient in a position of comfort:
  - Have uninjured patients lie down on their back with their feet about a foot higher than their head (Figure 7-1).
  - Keep injured patients comfortable in the position in which they are found.

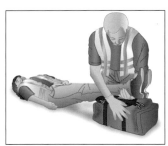

FIGURE 7-1

- Keep the patient warm and comfortable, perhaps by covering the patient with a blanket to help regulate body temperature.
- If the patient is wearing tight clothing, loosen it.
- Never give the patient any food or drink, even if the patient complains of thirst.
- Place uninjured patients in the recovery position (Figure 7-2) to protect their airway if they vomit or are bleeding from the mouth.

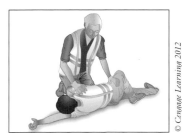

FIGURE 7-2

© Cengage Learning 2012

- If possible, treat the patient's injuries while waiting for EMS to arrive.
- Give emergency oxygen if available and you are trained to do so (Figure 7-3).

Shock is hard to detect and diagnose. Always be ready to treat for shock in any emergency. Remember, shock can be life threatening.

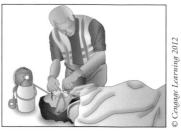

© Cengage Learning 2012

FIGURE 7-3

CHAPTER **8**

Jobsite Injuries

Jobsites can be dangerous places. The equipment and machinery found there present a whole host of hazards, and when we add in the human element the chances of an accident increase even more. The Occupational Safety and Health Administration (OSHA) has determined that about 90% of jobsite injuries occur within four areas, which OSHA calls the *Focus Four*:

© iStockphoto/Mel Stoutsenberger

- **Falls**—This includes, but is not limited to, falls from heights, falls off ladders or scaffolding, falls down stairs, falls caused by tripping over equipment, and so on.
- **Electrical**—This includes, but is not limited to, electrical shock from coming into contact with energized wires, equipment contacting overhead power lines, and so on.
- **Being struck**—This includes, but is not limited to, being struck by a moving vehicle or heavy equipment, being struck by falling or flying objects, and so on.
- **Being caught**—This includes, but is not limited to, being caught in a trench collapse, having fingers or other body parts caught in moving parts of equipment, and so on.

Electrical Working, 2010. Used under license from Shutterstock.com

As a first aid provider, you must be prepared to respond to a variety of jobsite injuries. Although that may seem over-whelming, if you follow what you have already learned, you can effectively manage most emergency situations.

Someonesfingers, 2010. Used under license from Shutterstock.com

Always remember the steps of "REACT." After any accident that causes an injury, make sure you take the time to:

Step 1—**R**ecognize a medical emergency

Step 2—**E**valuate the environment/Think safety first!

Step 3—**A**ssess responsiveness

Step 4—**C**all Emergency Medical Services (EMS)

Step 5—**T**ake action

It is now time to take action. With all jobsite injuries, keep in mind the three Cs:

- **Calm**—Keeping the patient calm will make your job easier.
- **Comfort**—Keeping the patient in a position of comfort is essential to minimizing further damage.
- **Care**—Providing basic first aid care to the patient until EMS arrives or the patient seeks other medical attention can save a life.

Major impact injuries will be discussed first, followed by minor impact injuries.

MAJOR IMPACT INJURIES

Injuries that result from a major physical force of energy against the body are referred to as major impact injuries. These injuries often occur on the jobsite as a result of falls, vehicle accidents, or being struck by falling or flying material. They should be considered very serious. If you encounter a patient with a major impact injury, always call EMS for advanced medical care, even if the patient is responsive.

Spinal Injuries

Signs and symptoms of spinal injuries include the following:

- Loss of movement
- Loss of sensation, including the ability to feel heat, cold, and touch

- Loss of bowel or bladder control
- Pain or an intense stinging sensation caused by damage to the nerves
- Difficulty breathing, coughing, or clearing secretions

Your role as a first aid provider is to do the following:

- Follow the steps of "REACT"—call EMS.
- Keep the patient calm.
- Provide comfort.
- Provide care:
  - Monitor the patient's airway and breathing and be ready to provide CPR.
  - Keep the patient still. Hold the patient's head with both of your hands in the position found to prevent movement of the neck and spine until EMS arrives.
  - Control any bleeding.
  - Suspect shock and begin treatment for shock.

Broken and Dislocated Bones

Signs and symptoms of broken and dislocated bones include the following:

- Extreme pain
- Deformity
- Exposed bone

Your role as a first aid provider is to do the following:

- Follow the steps of "REACT"—call EMS.
- Keep the patient calm.
- Provide comfort.
- Provide care:
  - Do not move the patient unless absolutely necessary.
  - Control any bleeding.
  - Immobilize the injured area with gentle care. Fill in the gaps between the injured area and the ground. If available, use padding such as a blanket or coat. Never try to realign the bone or shove a bone that is sticking out back in. If you have been trained how to splint and EMS is not readily accessible, apply a splint to the area above and below the injured area.

- Apply ice packs to the injured area to help limit swelling and alleviate pain until EMS arrives. It is not recommended that you apply ice directly to the skin; rather, wrap the ice in a cloth, towel, or clothing.
- Suspect shock and begin treatment for shock.

Head Injuries

Signs and symptoms of head injuries include the following:

- Unequal size of pupils
- Convulsions or vomiting
- Deformed features around the head and face
- Fluid draining from the patient's nose, mouth, or ears (may be clear or bloody)
- Impaired senses
- Inability to move one or more limbs
- Loss of consciousness
- Severe headache
- Slurred speech or impaired vision

Your role as a first aid provider is to do the following:

- Follow the steps of "REACT"—call EMS.
- Keep the patient calm.
- Provide comfort.
- Provide care:
  - Monitor the patient's airway and breathing and be ready to provide CPR.
  - Stabilize the patient's head and neck by placing your hands on both sides of the patient's head, keeping the head in line with the spine and preventing movement until you can turn the patient over to EMS.
- Control any bleeding.
- If you suspect a skull fracture, do not apply direct pressure to the bleeding site, and do not remove any debris from the wound. Cover the wound with sterile gauze dressing.
- Apply ice packs to the injured area to help limit swelling and alleviate pain until EMS arrives. It is not recommended that you apply ice directly to the skin; rather, wrap the ice in a cloth, towel, or clothing.
- Suspect shock and begin treatment for shock.

Impalement Injuries

Signs and symptoms of impalement injuries include the following:

- Object sticking out of the patient's body

Your role as a first aid provider is to do the following:

- Follow the steps of "REACT"—call EMS.
- Keep the patient calm.
- Provide comfort.
- Provide care:
  - Monitor the patient's airway and breathing and be ready to provide CPR.
  - Do not remove the impaled object. Impaled objects create a puncture wound and then put pressure on that same wound to assist in controlling bleeding. Removing the object may increase bleeding or cause further internal injury.
  - Control any bleeding without moving the impaled object.
  - Stabilize the object to prevent further damage, but never move the object.
  - Suspect shock and begin treatment for shock.

Chest Injuries

Signs and symptoms of chest injuries include the following:

- Deformed chest
- Fainting
- Abdominal pain
- Difficulty breathing
- Difficulty swallowing

Your role as a first aid provider is to do the following:

- Follow the steps of "REACT"—call EMS.
- Keep the patient calm.
- Provide comfort.
- Provide care:
  - Monitor the patient's airway and breathing and be ready to provide CPR.
  - Control any bleeding.
  - Keep the patient comfortable while you wait for EMS to arrive.
  - Suspect shock and begin treatment for shock.

Abdominal Injuries

Signs and symptoms of abdominal injuries include the following:

- Severe pain or tenderness in the area
- Holding or protecting the injured area
- Bruised, swollen, or rigid abdomen
- Rapid, shallow breathing
- Nausea or vomiting

Your role as a first aid provider is to do the following:

- Follow the steps of "REACT"—call EMS.
- Keep the patient calm.
- Provide comfort.
- Provide care:
  - Monitor the patient's airway and breathing and be ready to provide CPR.
  - For an open abdominal wound, cover the wound with a sterile dressing. Do not apply direct pressure.
  - Keep the patient comfortable while you wait for EMS to arrive.
  - Suspect shock and begin treatment for shock.

Amputations

Signs and symptoms of amputations include the following:

© iStockphoto/Nikki Bidgood

- Severed body part

Your role as a first aid provider is to do the following:

- Follow the steps of "REACT"— call EMS.
- Keep the patient calm.
- Provide comfort.
- Provide care:
  - Monitor the patient's airway and breathing and be ready to provide CPR.
  - Control bleeding caused by the amputation.
  - Take the severed body part and wrap it in a dry sterile dressing. Do not wash or attempt to otherwise clean the severed body part.

- Place the severed body part in a plastic bag and seal the bag.
- Take the sealed plastic bag and place it in a container on top of ice. Do not let the severed body part touch the ice directly. In addition, do not surround the severed body part with ice.
- Keep the patient comfortable while you wait for EMS to arrive.
- Suspect shock and begin treatment for shock.
- Turn over the severed body part to EMS when they arrive.

Severe Burns

Signs and symptoms of severe burns include the following:

Regenerate Burnt, 2010.
Used under license from
Shutterstock.com

- Charred black or dry white area
- Severe pain
- Difficulty breathing

Your role as a first aid provider is to do the following:

- Follow the steps of "REACT"—call EMS.
- Keep the patient calm.
- Provide comfort.
- Provide care:
  - Make sure the patient is no longer in contact with the material that caused the burn. This could include an electrical power line, fire, or chemical.
  - Monitor the patient's airway and breathing and be ready to provide CPR.
  - Make sure not to remove the patient's burned clothing.
  - Do not immerse large severe burns in cold water.
  - Raise the burned body part above the patient's heart level, when possible.
  - Use a cool, moist, sterile bandage; clean, moist cloth; or moist towel to cover the area of the burn.
  - Keep the patient comfortable while you wait for EMS to arrive.
  - Suspect shock and begin treatment for shock.

MINOR IMPACT INJURIES

Injuries that result from a less severe physical force of energy against the body are referred to as minor impact injuries. Although minor impact injuries may be painful, they are usually much less severe injuries and are not immediately life-threatening conditions. As a first aid provider, you may need to provide more assistance rather than actual care to the patient with a minor impact injury. The following are some examples of minor impact injuries and the assistance or care that should be given to the patient.

Sprains

Signs and symptoms of sprains include the following:

- Pain
- Swelling
- Bruising

Your role as a first aid provider is to do the following:

- Follow the steps of "REACT."
- Keep the patient calm.
- Provide comfort.
- Provide care:
  - Rest the injured area.
  - Support and do not move the injured area.
  - Place ice or a cold compress on the injured area.
  - Compress an injured extremity with roller gauze.
  - Elevate the injured extremity; a sling may be used for arm injuries.
  - Monitor and be ready to provide emergency care and call EMS if the patient develops a life-threatening condition.
  - If symptoms persist, encourage the patient to seek medical attention.

Eye Injuries

Signs and symptoms of eye injuries include the following:

- Pain
- Swelling

- Bruising
- Irritation to the eye

Your role as a first aid provider is to do the following:

- Follow the steps of "REACT."
- Keep the patient calm.
- Provide comfort.
- Provide care:
  - For eye injuries that are bleeding or leaking fluid, call EMS.
  - Put a cold compress over the eye for 10 to 15 minutes. Do not put pressure on the eye. If the patient is wearing contacts, do not remove them.
  - If there is dirt or small particles in the eye, gently flush the eye with water.
  - Monitor and be ready to provide emergency care and call EMS if the patient develops a life-threatening condition.
  - If symptoms persist, encourage the patient to seek medical attention.

Ear Injuries

Signs and symptoms of ear injuries include the following:

- Pain
- Swelling
- Bruising
- Full or partial loss of hearing

Your role as a first aid provider is to do the following:

- Follow the steps of "REACT."
- Keep the patient calm.
- Provide comfort.
- Provide care:
  - For ear injuries that are oozing clear fluid or watery blood, call EMS.
  - Use a sterile dressing and place it loosely over the ear. Do not apply pressure or plug the ear closed.
  - Monitor the patient and be ready to provide emergency care and call EMS if the patient develops a life-threatening condition.
  - If symptoms persist, encourage the patient to seek medical attention.

Knocked Out Teeth

Signs and symptoms of knocked out teeth include the following:

- Missing tooth
- Pain
- Bleeding

Your role as a first aid provider is to do the following:

- Follow the steps of "REACT."
- Keep the patient calm.
- Provide comfort.
- Provide care:
  - Control the bleeding by having the patient bite on a sterile gauze pad over the tooth socket for about 20 minutes.
  - Save the tooth and place it in a container of milk. Do not clean the tooth.
  - Have the patient see the dentist immediately to see if the tooth can be saved.
  - Monitor and be ready to provide emergency care and call EMS if the patient develops a life-threatening condition.

Minor Burns

Signs and symptoms of minor burns include the following:

- Red skin, with swelling, and pain
- Blisters
- Splotchy appearance

Your role as a first aid provider is to do the following:

© iStockphoto/bojan fatur

- Follow the steps of "REACT."
- Keep the patient calm.
- Provide comfort.
- Provide care:
  - If the burned area is large or if the burn is on the hands, feet, face, groin, or buttocks, treat it as a severe burn and call EMS immediately.

- Hold the burned area under cool (not cold) running water until the pain reduces. If you do not have access to running water, use a cold compress to cool the burn. Do not put ice on the burn as it can cause increased pain by cooling the burn too quickly.
- Use a sterile gauze bandage to cover the burn. Wrap the sterile gauze over the burn loosely to avoid setting weight or pressure on the burned area.
- Do not put any oil-based product on the burned area. There are many water-based burn creams and gels available to help reduce the pain.
- Monitor and be ready to provide emergency care and call EMS if the patient develops a life-threatening condition.
- If symptoms persist, encourage the patient to seek medical attention.

Puncture Wounds

Signs and symptoms of puncture wounds include the following:

- Evidence of impalement or puncture
- Pain
- Bleeding

Your role as a first aid provider is to do the following:

- Follow the steps of "REACT."
- Keep the patient calm.
- Provide comfort.
- Provide care:
  - Irrigate the wound with cool running water.
  - Using tweezers or a sterile gauze pad, attempt to remove any small particles from around the wound. Do not remove impaled objects.
  - Press softly on the outer area of the wound to promote bleeding.
  - Use a sterile gauze pad to dry the area and cover the wound with a sterile dressing.

- Monitor and be ready to provide emergency care and call EMS if the patient develops a life-threatening condition.
- Encourage the patient to seek medical attention.

Spider or Animal Bites

Signs and symptoms of spider or animal bites include the following:

- Bite marks
- Pain
- Bleeding
- Swelling

Your role as a first aid provider is to do the following:

- Follow the steps of "REACT."
- Keep the patient calm.
- Provide comfort.
- Provide care:
  - If the patient has difficulty breathing or you suspect an allergic reaction, call EMS immediately.
  - Monitor and be ready to provide emergency care and call EMS if the patient develops a life-threatening condition.
- Keep the bite area below the patient's heart level.
- Wash the area with soap and water.
- Cover the area with a sterile bandage and dressing.
- Use a cold compress on the bite area.
- Seek medical attention.

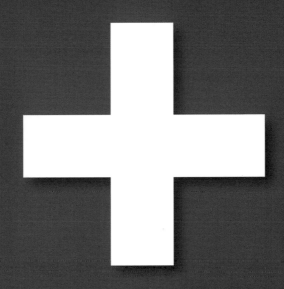

CHAPTER **9**

Jobsite Illnesses

There are numerous types of jobsite illnesses that have a detrimental effect on workers every day. As a first aid provider, you are not expected to be able to diagnose these illnesses or treat all of the signs and symptoms they present. Instead, your role regarding jobsite illnesses is primarily to recognize the emergency signs of illnesses.

EMERGENCY SIGNS OF ILLNESSES

- Difficulty breathing or shortness of breath
- Pain or pressure in the chest or abdomen
- Sudden dizziness or confusion
- Severe or persistent vomiting
- Flu-like symptoms that improve but then return with fever and a worsening cough

If one of these emergency signs of illness is present, you should act quickly and call EMS. This is the best way to help a patient who is showing one of these signs. You should always remember the steps of "REACT."

You should not put yourself in danger if the cause of the jobsite illness is environmental hazards such as exposure to chemicals. If you do not know the cause of an illness, proceed cautiously. As a first aid provider, you should take time to:

- **R**ecognize a medical emergency
- **E**valuate the environment/Think safety first!
- **A**ccess responsiveness
- **C**all Emergency Medical Services (EMS)
- **T**ake action

If an emergency warning sign is present, call EMS immediately.

As with jobsite injuries, you should keep the 3 Cs in mind when you respond to a jobsite illness:

- **Calm.** Keeping the patient calm will make your job as a first aid provider easier.
- **Comfort.** Keeping the patient in a position of comfort is essential to minimizing further damage.

- **Care.** Providing basic first aid care to the patient until EMS arrives or the patient seeks other medical attention can save a life.

Difficulty Breathing or Shortness of Breath

Signs and symptoms of difficulty breathing or shortness of breath include the following:

- Increased effort and rate of breathing
- Gasping or unable to catch breath
- Wheezing or gurgling with breathing
- Dizziness or lightheadedness

Your role as a first aid provider is to do the following:

- Follow the steps of "REACT"—call EMS.
- Keep the patient calm.
- Provide comfort.
- Provide care.
  - Monitor the patient's airway and breathing and be ready to provide CPR.
  - Permit the patient to find a comfortable position that allows for breathing.
  - If you are trained to do so, and it is available, provide the patient with emergency oxygen.
  - Assist the patient to take any prescribed medication, such as an inhaler for asthma or an epinephrine auto-injector for a severe allergic reaction.
  - Suspect shock and begin treatment for shock.

Pain or Pressure in the Chest

Signs and symptoms of pain or pressure in the chest include the following:

- Chest pain or pressure
- Lightheadedness
- Shortness of breath
- Nausea
- Pale, cool, and/or clammy skin

Your role as a first aid provider is to do the following:

- Follow the steps of "REACT"—call EMS.
- Keep the patient calm.
- Provide comfort.
- Provide care:
  - Monitor the patient's airway and breathing and be ready to provide CPR.
  - Permit the patient to find a comfortable position for breathing.
  - If you are trained to do so, and it is available, provide the patient with emergency oxygen.
  - Assist the patient to take any prescribed medication, such as nitroglycerin for angina.
  - Suspect shock and begin treatment for shock.

Pain or Pressure in the Abdomen

Signs and symptoms of pain or pressure in the abdomen include the following:

- Patient slumped over and grabbing stomach
- Pain that does not subside
- Lightheadedness
- Shortness of breath
- Nausea

Your role as a first aid provider is to do the following:

- Follow the steps of "REACT"—call EMS.
- Keep the patient calm.
- Provide comfort.
- Provide care:
  - Monitor the patient's airway and breathing and be ready to provide CPR.
  - Permit the patient to find a comfortable position.
  - If you are trained to do so, and it is available, provide the patient with emergency oxygen.
  - Suspect shock and begin treatment for shock.

Sudden Dizziness or Confusion

Signs and symptoms of sudden dizziness or confusion include the following:

- Diminished mental state
- Nausea
- Walking around without a purpose
- Unable to complete simple tasks

Your role as a first aid provider is to do the following:

- Follow the steps of "REACT"—call EMS.
- Keep the patient calm.
- Provide comfort.
- Provide care:
  - Monitor the patient's airway and breathing and be ready to provide CPR.
  - Permit the patient to find a comfortable position.
  - If you are trained to do so, and it is available, provide the patient with emergency oxygen.
  - Suspect shock and begin treatment for shock.

Severe or Persistent Vomiting

Signs and symptoms of severe or persistent vomiting include the following:

- Vomiting that does not subside
- Severe abdominal pain
- Confusion

Your role as a first aid provider is to do the following:

- Follow the steps of "REACT"—call EMS.
- Keep the patient calm.
- Provide comfort.
- Provide care:
  - Monitor the patient's airway and breathing and be ready to provide CPR.
  - Permit the patient to find a comfortable position.

- If you are trained to do so, and it is available, provide the patient with emergency oxygen.
- Suspect shock and begin treatment for shock.

Flu-like Symptoms Improve but Then Return with Fever and Worsening Cough

Signs and symptoms of flu-like symptoms that return and worsen include the following:

- Varying flu symptoms
- Fever
- Worsening cough

Your role as a first aid provider is to do the following:

- Follow the steps of "REACT"—call EMS.
- Keep the patient calm.
- Provide comfort.
- Provide care:
  - Monitor the patient's airway and breathing and be ready to provide CPR.
  - Permit the patient to find a comfortable position.
  - If you are trained to do so, and it is available, provide the patient with emergency oxygen.
  - Suspect shock and begin treatment for shock.

SPECIFIC JOBSITE ILLNESSES

Severe Heat Exposure

Signs and symptoms of severe heat exposure include the following:

- Heavy sweating
- Pale, cool, moist skin or very warm skin
- Exhaustion, dizziness, and/or weakness
- Muscle cramps
- Reduced mental state

© iStockphoto/BartCo

Your role as a first aid provider is to do the following:

- Follow the steps of "REACT"—call EMS.
- Keep the patient calm.

- Provide comfort.
- Provide care:
  - Monitor the patient's airway and breathing and be ready to provide CPR.
  - Remove the patient from the hot environment to a place which is cooler.
  - Assist the patient in lying down, and elevate the patient's legs.
  - If the patient has an alert mental state, give water to help hydrate.
  - Apply cold water, wet towels, or cold compresses to rapidly cool the patient's body.
  - If the patient does not respond favorably to these treatments or the patient's condition worsens, call EMS.

Extreme Cold Exposure

Signs and symptoms of extreme cold exposure include the following:

- Cool or cold skin temperature
- Wet or damp clothing
- Uncontrollable shivering
- Reduced mental state
- Trouble walking, talking, or completing simple tasks

Your role as a first aid provider is to do the following:

- Follow the steps of "REACT"—call EMS.
- Keep the patient calm.
- Provide comfort.
- Provide care:
  - Monitor the patient's airway and breathing and be ready to provide CPR.
  - Remove the patient from the cold environment to a place which is warmer.
  - Protect the patient from losing further body heat by assisting the patient in removing wet clothing, patting the skin dry, and covering the patient with blankets.
  - Do not give the patient anything to eat or drink.
  - Do not rub down any part of the body or allow the patient to move around too much.

- Do not immerse the patient in hot water or use direct heat because rapid warming can cause heart troubles.
- If the patient does not respond favorably to these treatments or the condition worsens, call EMS.

Allergic Reaction

Signs and symptoms of allergic reaction include the following:

- Difficulty breathing, wheezing, or tightness in throat and/or chest
- Swelling of the face or neck, puffy eyes
- Anxiety or agitation
- Raised lumps or hives on the skin
- Nausea and/or vomiting
- Changing levels of responsiveness

Image Copyright Banks Photos, 2009. Used under license from

Your role as a first aid provider is to do the following:

- Follow the steps of "REACT"—call EMS.
- Keep the patient calm.
- Provide comfort.
- Provide care:
  - Monitor the patient's airway and breathing and be ready to provide CPR.
  - Permit the patient to find a comfortable position for breathing.
  - If you are trained to do so, and it is available, provide the patient with emergency oxygen.
  - Assist the patient to take any prescribed medication, such as an epinephrine auto-injector for a severe allergic reaction.
  - Suspect shock and begin treatment for shock.

Diabetic Emergencies

Signs and symptoms of diabetic emergencies include the following:

- Sudden dizziness and/or shakiness
- Headache and confusion

- Pale skin and sweating
- Awkward or jumpy movements
- Changing level of responsiveness
- Breath smells fruity
- Dry mouth and thirsty

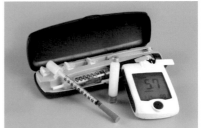

Image Copyright Givaga, 2012. Used under license from Shutterstock.com

Your role as a first aid provider is to do the following:

- Follow the steps of "REACT"—call EMS.
- Keep the patient calm.
- Provide comfort.
- Provide care:
  - Monitor the patient's airway and breathing and be ready to provide CPR.
  - Confirm if the patient has diabetes by looking for a medical alert tag or communicating with the patient if responsive.
  - Permit the patient to find a comfortable position for breathing.
  - Assist the patient in taking sugar (glucose tablets, fruit juice, sugar packets, and hard candy).
  - If the patient does not respond favorably to these treatments or the condition worsens, call EMS.

Heart Attack

Signs and symptoms of heart attack include the following:

- Feeling of impending gloom
- Sweating
- Chest pain and pressure
- Jaw pain
- Nausea and/or vomiting
- Back pain
- Pain spreading to the shoulders and arms
- Indigestion

Your role as a first aid provider is to do the following:

- Follow the steps of "REACT"—call EMS.
- Keep the patient calm.

- Provide comfort.
- Provide care:
  - Monitor the patient's airway and breathing and be ready to provide CPR.
  - Assist the patient in finding a position of comfort.
  - Loosen any clothing that may be constricting.
  - Do not give the patient anything to eat or drink.
  - Allow the patient to chew or swallow one aspirin, unless the patient is allergic.
  - Assist the patient to take any prescribed medication, such as nitroglycerin for angina.
  - Suspect shock and begin treatment for shock.

Asthma

Signs and symptoms of asthma include the following:

- Difficulty breathing with long, difficult exhalation
- Increased breathing rate
- High-pitched, hacky-type cough
- Patient sitting forward with hands on knees or chair
- Anxiety or agitation
- Blue nail beds or lips

Image Copyright RonaldSumners, 2012. Used under license from Shutterstock.com

Your role as a first aid provider is to do the following:

- Follow the steps of "REACT"—call EMS.
- Keep the patient calm.
- Provide comfort.
- Provide care:
  - Monitor the patient's airway and breathing and be ready to provide CPR.
  - Permit the patient to find a comfortable position for breathing.
  - If you are trained to do so, and it is available, provide the patient with emergency oxygen.

- Assist the patient to take any prescribed medication, such as a rescue inhaler.
- If the patient does not respond favorably to these treatments or the condition worsens, call EMS.

Poisoning

Signs and symptoms of poisoning include the following:

- Open containers or bottles nearby
- Burns on the mouth and/or hands
- Changing level of responsiveness
- Nausea and/or vomiting
- Chest pain
- Abdominal pain or cramping
- Headache, lightheadedness, and/or dizziness

Image Copyright Chictype, 2007. Used under license from Shutterstock.com

Your role as a first aid provider is to do the following:

- Follow the steps of "REACT"—call EMS.
- Keep the patient calm.
- Provide comfort.
- Provide care:
  - Monitor the patient's airway and breathing and be ready to provide CPR.
  - Assist the patient in moving away from the area if the poison is still in the vicinity; however, only enter the area if it is safe to do so.
  - Loosen any tight and constricting clothing.
  - If the patient is responsive, call the Poison Control Center (1-800-222-1222) and follow their directions.
  - If the patient is unresponsive, call EMS immediately.
  - Suspect shock and begin treatment for shock.

Seizure

Signs and symptoms of spinal injuries include the following:

- Sudden stiffening and falling to the ground
- Quick jerking movements
- Loss of bowel or bladder control
- Nausea and/or vomiting
- Difficulty breathing or hissing breath

Your role as a first aid provider is to do the following:

- Follow the steps of "REACT"—call EMS.
- Keep the patient calm.
- Provide comfort.
- Provide care:
  - Monitor the patient's airway and breathing and be ready to provide CPR.
  - Do not attempt to stop the patient's movements or otherwise restrain the patient.
  - Do not place any objects in the patient's mouth.
  - Move objects away from the patient to prevent further injury.
  - If the patient vomits, use the log roll to help clear fluid from the patient's mouth (Figure 9-1).
  - Suspect shock and begin treatment for shock.

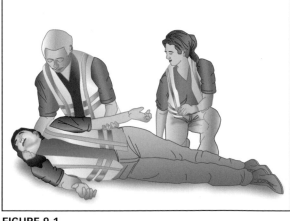

© Cengage Learning 2012

FIGURE 9-1

Stroke

Signs and symptoms of stroke include the following:

- Collapse or falling to the ground
- Difficulty in breathing or snoring while breathing
- Loss of bowel or bladder control
- Convulsions or seizures
- Difficulty with speech or vision
- Numbness or weakness on one side of the body
- Changing level of responsiveness

Your role as a first aid provider is to do the following:

- Follow the steps of "REACT"—call EMS.
- Keep the patient calm.
- Provide comfort.
- Provide care:
  - Monitor the patient's airway and breathing and be ready to provide CPR.
  - Permit the patient to find a comfortable position.
  - If unresponsive but breathing, place the patient in the recovery position (as shown in Chapter 3) to protect the airway.
  - Provide emotional support to the responsive patient.
  - Suspect shock and begin treatment for shock.

Fainting

Signs and symptoms of fainting include the following:

- Collapse or falling to the ground
- Changing level of responsiveness

Your role as a first aid provider is to do the following:

- Follow the steps of "REACT."
- Keep the patient calm.
- Provide comfort.
- Provide care:
  - Monitor the patient's airway and breathing and be ready to provide CPR.
  - Assist the patient to lie on a firm, comfortable surface, with the patient's back flat against the surface.

- Elevate the patient's legs 8 to 12 inches off the ground.
- Loosen any tight or constricting clothing.
- Check for injuries and treat those injuries.
- Call EMS if:
  - The patient does not regain responsiveness soon after fainting.
  - The patient faints repeatedly.
  - The patient has heart disease.
  - The patient is pregnant.

Drug or Alcohol Overdose

Signs and symptoms of drug or alcohol ovedose can vary greatly depending on the drug or alcohol to which the patient was exposed. Some general signs and symptoms are:

- Changing level of responsiveness
- Lowered blood pressure
- Glazed eye appearance

Your role as a first aid provider is to do the following:

- Follow the steps of "REACT"—call EMS.
- Keep the patient calm.
- Provide comfort.
- Provide care:
  - Monitor the patient's airway and breathing and be ready to provide CPR.
  - If the patient is responsive:
    - Stay with the patient and provide protection from injury.
    - Do not allow the victim to lie down on his or her back.
    - If the patient took a drug, find out which drug, call the Poison Control Center (1-800-222-1222), and follow their directions.

Emergency Preparedness

Employees at all jobsites should take steps to prepare for an emergency. This guide has presented a lot of information you can use as a first aid provider for medical emergencies. However, many other emergencies should be planned for, so employers and their employees can safely respond to those emergencies at the jobsite. There are many resources available to assist employers in preparing an emergency action plan. The Occupational Safety & Health Administration's website on Safety and Health Topics (http:www.osha.gov/SLTC) is a good source for employers.

EMERGENCY ACTION PLAN CHECKLIST

Table 10-1

Emergency Action Plan Checklist

GENERAL ISSUES

Does the plan consider all natural or man-made emergencies that could disrupt your workplace?	Common sources of emergencies identified in emergency action plans include fires, explosions, floods, hurricanes, tornadoes, toxic material releases, radiological and biological accidents, civil disturbances, and workplace violence.
Does the plan consider all potential internal sources of emergencies that could disrupt your workplace?	Conduct a hazard assessment of the workplace to identify any physical or chemical hazards that may exist and could cause an emergency.
Does the plan consider the impact of these internal and external emergencies on the workplace's operations and is the response tailored to the workplace?	Brainstorm worst-case scenarios by asking yourself what you would do and what would be the likely impact on your operation, then devise appropriate responses.
Does the plan contain a list of key personnel with contact information as well as contact information for local emergency responders, agencies, and contractors?	Keep your list of key contacts current and make provisions for an emergency communications system such as a cellular phone, a portable radio unit, or other means so that contact with local law enforcement, the fire department, and others can be swift.

GENERAL ISSUES	
Does the plan contain the names, titles, departments, and telephone numbers of individuals to contact for additional information or an explanation of duties and responsibilities under the plan?	List names and contact information for individuals responsible for implementation of the plan.
Does the plan address how rescue operations will be performed?	Unless you are a large employer handling hazardous materials and processes or have employees regularly working in hazardous situations, you will probably choose to rely on local public resources, such as the fire department, who are trained, equipped, and certified to conduct rescues. Make sure any external department or agency identified in your plan is prepared to respond as outlined in your plan. Untrained individuals may endanger themselves and those they are trying to rescue.
Does the plan address how medical assistance will be provided?	Most small employers do not have a formal internal medical program and make arrangements with medical clinics or facilities close by to handle emergencies. If an infirmary, clinic, or hospital is not close to your workplace, ensure that onsite person(s) have adequate training in first aid. The American Red Cross, some insurance providers, local safety councils, fire departments, or other resources may be able to provide this training. Treatment of a serious injury should begin within 3 to 4 minutes of the accident. Consult with a physician to order appropriate first-aid supplies for emergencies. Establish a relationship with a local ambulance service so transportation is readily available for emergencies.

(Continued)

Table 10-1 (cont.)

GENERAL ISSUES

Does the plan identify how or where personal information on employees can be obtained in an emergency?	In the event of an emergency, it could be important to have ready access to important personal information about your employees. This includes their home telephone numbers, the names and telephone numbers of their next of kin, and medical information.

EVACUATION POLICY AND PROCEDURE

Does the plan identify the conditions under which an evacuation would be necessary?	The plan should identify the different types of situations that will require an evacuation of the workplace. This might include a fire, earthquake, or chemical spill. The extent of evacuation may be different for different types of hazards.
Does the plan identify a clear chain of command and designate a person authorized to order an evacuation or shutdown of operations?	It is common practice to select a responsible individual to lead and coordinate your emergency plan and evacuation. It is critical that employees know who the coordinator is and understand that this person has the authority to make decisions during emergencies. The coordinator should be responsible for assessing the situation to determine whether an emergency exists requiring activation of the emergency procedures, overseeing emergency procedures, notifying and coordinating with outside emergency services, and directing shutdown of utilities or plant operations if necessary.
Does the plan address the types of actions expected of different employees for the various types of potential emergencies?	The plan may specify different actions for employees depending on the emergency. For example, employers may want to have employees assemble in one area of the workplace if it is threatened by a tornado or earthquake but evacuate to an exterior location during a fire.

EVACUATION POLICY AND PROCEDURE	
Does the plan designate who, if anyone, will stay to shut down critical operations during an evacuation?	You may want to include in your plan locations where utilities (such as electrical and gas utilities) can be shut down for all or part of the facility. All individuals remaining behind to shut down critical systems or utilities must be capable of recognizing when to abandon the operation or task and evacuate themselves.
Does the plan outline specific evacuation routes and exits and are these posted in the workplace where they are easily accessible to all employees?	Most employers create maps from floor diagrams with arrows that designate the exit route assignments. These maps should include locations of exits, assembly points, and equipment (such as fire extinguishers, first aid kits, spill kits) that may be needed in an emergency. Exit routes should be clearly marked and well lit, wide enough to accommodate the number of evacuating personnel, unobstructed and clear of debris at all times, and unlikely to expose evacuating personnel to additional hazards.
Does the plan address procedures for assisting people during evacuations, particularly those with disabilities or who do not speak English?	Many employers designate individuals as evacuation wardens to help move employees from danger to safe areas during an emergency. Generally, one warden for every 20 employees should be adequate, and the appropriate number of wardens should be available at all times during working hours. Wardens may be responsible for checking offices and bathrooms before being the last person to exit an area as well as ensuring that fire doors are closed when exiting. Employees designated to assist in emergency evacuation procedures should be trained in the complete workplace layout and various alternative escape routes. Employees designated to assist

(*Continued*)

Table 10-1 (cont.)

EVACUATION POLICY AND PROCEDURE

	in emergencies should be made aware of employees with special needs (who may require extra assistance during an evacuation), how to use the buddy system, and any hazardous areas to avoid during an emergency evacuation.
Does the plan identify one or more assembly areas (as necessary for different types of emergencies) where employees will gather and a method for accounting for all employees?	Accounting for all employees following an evacuation is critical. Confusion in the assembly areas can lead to delays in rescuing anyone trapped in the building, or unnecessary and dangerous search-and-rescue operations. To ensure the fastest, most accurate accounting of your employees, consider taking a head count after the evacuation. The names and last known locations of anyone not accounted for should be passed on to the official in charge.
Does the plan address how visitors will be assisted in evacuation and accounted for?	Some employers have all visitors and contractors sign in when entering the workplace. The hosts and/or area wardens, if established, are often tasked with assisting these individuals to evacuate safely.

REPORTING EMERGENCIES AND ALERTING EMPLOYEES IN AN EMERGENCY

Does the plan identify a preferred method for reporting fires and other emergencies?	Dialing 911 is a common method for reporting emergencies if external responders are utilized. Internal numbers may be used as well. Internal numbers are sometimes connected to intercom systems so that coded announcements may be made. In some cases employees are requested to activate manual pull stations or other alarm systems.

REPORTING EMERGENCIES AND ALERTING EMPLOYEES IN AN EMERGENCY	
Does the plan describe the method to be used to alert employees, including disabled workers, to evacuate or take other action?	Make sure alarms are distinctive and recognized by all employees as a signal to evacuate the work area or perform other actions identified in your plan. Sequences of horn blows or different types of alarms (bells, horns, etc.) can be used to signal different responses or actions from employees. Consider making available an emergency communications system, such as a public address system, for broadcasting emergency information to employees. Ideally alarms will be able to be heard, seen, or otherwise perceived by everyone in the workplace including those that may be blind or deaf. Otherwise floor wardens or others must be tasked with ensuring all employees are notified. You might want to consider providing an auxiliary power supply in the event of an electrical failure.
EMPLOYEE TRAINING AND DRILLS	
Does the plan identify how and when employees will be trained so that they understand the types of emergencies that may occur, their responsibilities, and actions as outlined in the plan?	Training should be offered to employees when you develop your initial plan and when new employees are hired. Employees should be retrained when your plan changes due to a change in the layout or design of the facility; when new equipment, hazardous materials, or processes are introduced that affect evacuation route; or when new types of hazards are introduced that require special actions. General training for your employees should address the following: • Individual roles and responsibilities; • Threats, hazards, and protective actions; • Notification, warning, and communications procedures; • Emergency response procedures;

(*Continued*)

Table 10-1 (cont.)

EMPLOYEE TRAINING AND DRILLS

	• Evacuation, shelter, and accountability procedures; • Location and use of common emergency equipment; and • Emergency shutdown procedures. You may also need to provide additional training to your employees (i.e., first aid procedures, portable fire extinguisher use, etc.) depending on the responsibilities allocated to employees in your plan.
Does the plan address how and when retraining will be conducted?	If training is not reinforced it will be forgotten. Consider retraining employees annually.
Does the plan address if and how often drills will be conducted?	Once you have reviewed your emergency action plan with your employees and everyone has had the proper training, it is a good idea to hold practice drills as often as necessary to keep employees prepared. Include outside resources such as fire and police departments when possible. After each drill, gather management and employees to evaluate the effectiveness of the drill. Identify the strengths and weaknesses of your plan and work to improve it.

© Cengage Learning 2012

HOUSEHOLD EMERGENCY PLAN

Individuals should also create an emergency action plan for their homes. A key element to these plans is a Household Emergency Information Card. This card should communicate emergency phone numbers and elements of the emergency action plan. The card should be displayed in a prominent location and pointed out to any houseguests or visitors. Table 10-2 shows an example of a Household Emergency Information Card:

Table 10-2

Household Emergency Information Card

Emergency Number: _____

This phone number is:	Directions to this house:
This address is:	
Emergency Contact Numbers	**Household Emergency Action Plan**
Police Department:	In case of emergency, the family will meet here:
Fire Department:	
Ambulance:	or here:
Poison Control Center:	
Family Physician:	Emergency Contact Name:
Father Work Phone:	Phone Number:
Mother Work Phone:	Address:
Father Cell Phone:	
Mother Cell Phone:	
Trusted Neighbors Name:	
Trusted Neighbors Phone:	First Aid Kit Located:
Relative/Friend Phone:	
Relative/Friend Phone:	Plan for Pets:
Other Phone Numbers:	
Notes:	

© Cengage Learning 2012

FIRST AID KIT

A well-stocked first aid kit can help you, as a first aid provider, effectively respond to many types of emergencies. The first aid kit should be stored in a readily accessible location. Multiple first aid kits should be used at large jobsites. A good rule of thumb is to make sure the first aid kit can be reached within 5 minutes from all work areas of the jobsite.

You can purchase first aid kits at many stores and from safety supply companies, or you can assemble your own. The first aid kit should be designed and filled to allow you or other first aid providers to access supplies they need to respond to potential jobsite emergencies. First aid kits should also be maintained to keep an adequate stock of supplies.

First Aid Kit Supplies

A first-aid kit should include:

- Adhesive tape
- Antibiotic ointment
- Antiseptic spray or towelettes
- Bandages, including a roll of elastic wrap and bandage strips in assorted sizes
- Burn spray or cream
- Cold compresses (instant)
- Cotton-tipped applicators
- CPR mask or shield
- Disposable gloves, at least two pairs
- Emergency eye wash solution
- First aid blanket
- Gauze pads and roller gauze in assorted sizes
- Hydrocortisone cream
- Instant hand sanitizer
- Medications
- Red biohazard plastic bags for the disposal of contaminated materials
- Scissors
- Splinter-removing tool
- Sterile eye wash, such as a saline solution
- Thermometer
- Triangular bandage
- Tweezers

The items in a first aid kit should be individually packaged when multiple individuals will use the first aid kit. This will help prevent cross contamination.

VEHICLE EMERGENCY KIT

Many workers at jobsites work out of their vehicles. Each vehicle should have a first aid kit and a vehicle emergency kit (Figure 10-1). A vehicle emergency kit is designed to assist the driver and passengers of the vehicle if they experience an emergency that could leave them stranded for an extended period of time.

© iStockphoto/ra3rn

Vehicle Emergency Kit Supplies

FIGURE 10-1

The following are some recommended supplies for your vehicle emergency kit. Depending on where you work and travel, more supplies may be needed.

- Blankets and/or sleeping bag
- Candles and matches
- Cell phone and charger
- Emergency phone numbers, including contact information for local emergency services, emergency road service providers, and the regional poison control center
- Extra coats and other clothing
- Fire extinguisher
- First aid kit
- Instant tire inflation can
- Nonperishable food and water
- Portable radio
- Road flares and emergency beacons
- Small, waterproof flashlight and extra batteries
- Sunscreen

CHAPTER **11**

**OSHA Reporting
and Recordkeeping**

Regardless of the outcome of a jobsite emergency, employers have a requirement to report jobsite fatalities and catastrophes and record jobsite injuries and illnesses. This chapter provides information on the requirements for reporting and recording jobsite accidents, injuries, and illnesses.

JOBSITE FATALITIES AND CATASTROPHES

From 2005 through 2009 there was an average of over 5,300 work-related deaths in the United States. The Occupational Safety and Health Administration (OSHA) takes each and every one of these fatalities seriously and investigates them. OSHA requires all jobsite fatalities and catastrophes to be reported to them by the employer. The OSHA regulation 29 CFR 1904.39(a) states:

> Within eight (8) hours after the death of any employee from a work-related incident or the in-patient hospitalization of three or more employees as a result of a work-related incident, you must orally report the fatality/multiple hospitalization by telephone or in person to the Area Office of the Occupational Safety and Health Administration (OSHA), U.S. Department of Labor, that is nearest to the site of the incident. You may also use the OSHA toll-free central telephone number, 1-800-321-OSHA (1-800-321-6742).

RECORDING JOBSITE INJURIES AND ILLNESSES

Injuries and illnesses that occur at the workplace need to be recorded on the appropriate OSHA recordkeeping forms. Over the past couple of years, OSHA has instituted an emphasis program on recordkeeping to ensure employers are complying with regulations. OSHA provides employers with the following information on their recordkeeping requirements.

OSHA Publication 3169

Employers now have a new system for tracking workplace injuries and illnesses. OSHA's new recordkeeping log is easier to understand and to use. Written in plain language using a question-and-answer format, the revised recordkeeping rule answers questions about recording occupational injuries and illnesses and explains how to classify particular cases. Flowcharts and checklists make it easier to follow the recordkeeping requirements.

What Has Changed?

The new rule:

- Offers flexibility by letting employers computerize injury and illness records;
- Updates three recordkeeping forms:
  - OSHA Form 300 (Log of Work-Related Injuries and Illnesses); simplified and reformatted to fit legal size paper.
  - OSHA Form 301 (Injury and Illness Incident Report); includes more data about how the injury or illness occurred.
  - OSHA Form 300A (Summary of Work-Related Injuries and Illnesses); separate form created to make it easier to calculate incidence rates;
- Continues to exempt smaller employers (employers with 10 or fewer employees) from most requirements;
- Changes the exemptions for employers in service and retail industries;
- Clarifies the definition of work relationship, limiting the recording of pre-existing cases and adding new exceptions for some categories of injury and illness;
- Includes new definitions of medical treatment, first aid, and restricted work to simplify recording decisions;
- Eliminates different criteria for recording work-related injuries and work-related illnesses; one set of criteria will be used for both;
- Changes the recording of needlestick injuries and tuberculosis;
- Simplifies the counting of days away from work, restricted days, and job transfer;
- Improves employee involvement and provides employees and their representatives with access to the information; and
- Protects privacy for injured and ill workers.

Simplified, clearer definitions also make it easier for employers to determine which cases must be recorded. Posting an annual summary of workplace injuries and illnesses for a longer period of time improves employee access to information, and as employees learn how to report workplace injuries and illnesses, their involvement and participation increase.

Which Recordkeeping Requirements Apply to Me?

Reporting fatalities and catastrophes: All employers covered by the Occupational Safety and Health Act of 1970 (P.L. 91-596) must report to OSHA any workplace incident resulting in a fatality or the in-patient hospitalization of three or more employees within 8 hours. Keeping injury and illness records: If you had 10 or fewer employees during all of the last calendar year or your business is classified in a specific low-hazard retail, service, finance, insurance, or real estate industry, you do not have to keep injury and illness records unless the Bureau of Labor Statistics or OSHA informs you in writing that you must do so.

How Can I Tell if I Am Exempt?

OSHA uses the Standard Industrial Classification (SIC) Code to determine which establishments must keep records. You can search for SIC Codes by keywords or by four-digit SIC to retrieve descriptive information of specific SICs in OSHA's online North American Industry Classification System Search, available on OSHA's website at: http://www.osha.gov/oshstats/naics-manual.html. Establishments classified in the following SICs are exempt from most of the record-keeping requirements, regardless of size:

525 Hardware Stores
542 Meat and Fish Markets
544 Candy, Nut, and Confectionary Stores
545 Dairy Products Stores
546 Retail Bakeries
549 Miscellaneous Food Stores
551 New and Used Car Dealers
552 Used Car Dealers
554 Gasoline Service Stations
557 Motorcycle Dealers
56   Apparel and Accessory Stores
573 Radio, Television, and Computer Stores
58   Eating and Drinking Places
591 Drug Stores and Proprietary Stores
592 Liquor Stores
594 Miscellaneous Shopping Goods Stores
599 Retail Stores, Not Elsewhere Classified
60   Depository Institutions (Banks and Savings Institutions)

61   Nondepository Institutions (Credit Institutions)
62   Security and Commodity Brokers
63   Insurance Carriers
64   Insurance Agents, Brokers, and Services
653   Real Estate Agents and Managers
654   Title Abstract Offices
67   Holding and Other Investment Offices
722   Photographic Studios, Portrait
723   Beauty Shops
724   Barber Shops
725   Shoe Repair and Shoeshine Parlors
726   Funeral Service and Crematories
729   Miscellaneous Personal Services
731   Advertising Services
732   Credit Reporting and Collection Services
733   Mailing, Reproduction, and Stenographic Services
737   Computer and Data Processing Services
738   Miscellaneous Business Services
764   Reupholstery and Furniture Repair
78   Motion Picture
791   Dance Studios, Schools, and Halls
792   Producers, Orchestras, Entertainers
793   Bowling Centers
801   Offices and Clinics of Medical Doctors
802   Offices and Clinics of Dentists
803   Offices of Osteopathic Physicians
804   Offices of Other Health Practitioners
807   Medical and Dental Laboratories
809   Health and Allied Services, Not Elsewhere Classified
81   Legal Services
82   Educational Services (Schools, Colleges, Universities, and Libraries)
832   Individual and Family Services
835   Child Day Care Centers
839   Social Services, Not Elsewhere Classified
841   Museums and Art Galleries
86   Membership Organizations
87   Engineering, Accounting, Research, Management, and Related Services
899   Services, Not Elsewhere Classified

What Do I Have to Do If I Am Not Exempt?

Employers not exempt from OSHA's recordkeeping requirements must prepare and maintain records of work-related injuries and illnesses. You need to review Title 29 of the Code of Federal Regulations (CFR) Part 1904, "Recording and Reporting Occupational Injuries and Illnesses," to see exactly which cases to record.

- **Use the Log of Work-Related Injuries and Illnesses (*Form 300*)** to list injuries and illnesses and track days away from work, restricted, or transferred.
- **Use the Injury and Illness Report (*Form 301*)** to record supplementary information about recordable cases. You can use a workers' compensation or insurance form, if it contains the same information.
- **Use the Summary (*Form 300A*)** to show totals for the year in each category. The summary is posted from February 1 to April 30 of each year.

What Is So Important About Recordkeeping?

Recordkeeping is a critical part of an employer's safety and health efforts for several reasons:

- Keeping track of work-related injuries and illnesses can help you prevent them in the future.
- Using injury and illness data helps identify problem areas. The more you know, the better you can identify and correct hazardous workplace conditions.
- You can better administer company safety and health programs with accurate records.
- As employee awareness about injuries, illnesses, and hazards in the workplace improves, workers are more likely to follow safe work practices and report workplace hazards. OSHA compliance officers can rely on the data to help them properly identify and focus on injuries and illnesses in a particular area. The agency also asks about 80,000 establishments each year to report the data directly to OSHA, which uses the information as part of its site-specific inspection targeting program. The Bureau of Labor Statistics (BLS) also uses injury and illness records as the source data for the Annual Survey of Occupational Injuries and Illnesses that shows safety and health trends nationwide and industry wide.

How Can I Get More Information on Recordkeeping?

The full preamble and text of the new rule is available online. You can find it by searching the Index on OSHA's website at http://www.osha.gov. You can also receive a copy of the regulation from OSHA's Office of Publications, P.O. Box 37535, Washington, DC 20013-7535; phone (202) 693-1888. If your workplace is in a state operating under an OSHA-approved plan, state plan recordkeeping regulations, although similar to federal ones, may have some more stringent or supplemental requirements such as reporting fatalities and catastrophes. Industry exemptions may also differ. For further information and assistance, you may call OSHA at 1-800-321-OSHA. Teletypewriter (TTY) number is 1-877-889-5627. Also visit OSHA's website at http://www.osha.gov to get contact information for the following states: Alaska, Arizona, California, Hawaii, Indiana, Iowa, Kentucky, Maryland, Michigan, Minnesota, Nevada, New Mexico, North Carolina, Oregon, Puerto Rico, South Carolina, Tennessee, Utah, Vermont, Virginia, Virgin Islands, Washington, and Wyoming. In other states, contact the nearest OSHA Regional Office listed here and ask for the recordkeeping coordinator:

Atlanta .... (404) 562-2300
Boston .... (617) 565-9860
Chicago........................... (312) 353-2220
Dallas .... (214) 767-4731
Denver .... (303) 844-1600
Kansas City..................... (816) 426-5861
New York .... (212) 337-2378
Philadelphia................... (215) 861-4900
San Francisco................. (415) 975-4310
Seattle .... (206) 553-5930

CHAPTER 12

Accident and Incident Investigations

As presented in Chapter 11, recordkeeping is a critical part of an employer's safety and health efforts for several reasons:

- Keeping track of work-related injuries and illnesses can help the employer prevent them in the future.
- Using injury and illness data helps identify problem areas. The more employers know, the better they can identify and correct hazardous workplace conditions.
- Employers can better administer company safety and health programs with accurate records.

It is also critically important to investigate all accidents and incidents that occur at the jobsite. By investigating, an employer will be better equipped to find the root cause of the accident or incident. Understanding the root cause will assist employers as they make continued improvements to their safety and health programs.

ACCIDENT VS. INCIDENT

Investigating accidents is a no-brainer. Employers want to find out what happened, discover contributing factors, and determine the root cause so they can prevent a re-occurrence of the accident. But why do employers also investigate incidents? Let us start by defining each term:

- **Accident.** An accident is defined as an undesirable event that occurs unintentionally and usually results in harm, injury, damage, or loss.
- **Incident.** An incident is defined as an individual occurrence or event which is likely to take place again if all the elements leading up to the incident remain the same. An incident is also referred to as a near miss.

The main difference between an accident and an incident is that an accident results in a loss and an incident does not result in a loss. Many employers understand that the accident caused the injury. However, a similar event could occur but not result in an accident. As an example, a first aid provider responded to the scene of an accident where a forklift was overturned and crushed the leg of the operator. Once the paramedics left with the injured forklift operator, workers standing around the jobsite discussed what they had seen. A couple workers told stories of the same forklift operator driving the forklift too fast on the jobsite. Other workers told accounts of seeing the forklift operator driving the forklift and not wearing his seatbelt. One worker spoke about a time when the forklift almost rolled because of operator error.

© iStockphoto/chinmy

The employer wished they had heard that information prior to the accident. They could have re-trained the operator and taken other steps to prevent those incidents from re-occurring. Instead, they have an employee who lost his leg in an accident. If they had a system to investigate all incidents, they could have prevented a major accident.

ACCIDENT AND INCIDENT INVESTIGATIONS

When to Conduct the Investigation

It is important that any investigation occurs as soon as possible. The less time between an accident or incident and the investigation, the more accurate the information that can be obtained.

Goal of the Investigation

A key element to the investigation is to examine the causes and results of any accident or incident without prejudice. The investigator must begin the investigation with an open mind. No hypothesis should be made and any conclusion should be based on information that is known to be full and accurate.

It is helpful for the investigator to ask open-ended questions and not to put words into witnesses' mouths. The investigator should not blame people but rather emphasize the significance of seeking the reasons for the incident to prevent a re-occurrence.

It is far less useful to attempt to change people and their behaviors than it is to change their surroundings so that the consequences of a mistake on their part are either eradicated or reduced. Instead of an emotional discussion of patterns of behavior, a more constructive approach attempting to modify the environment is needed. It is more effective to adjust the circumstances producing a mistake than to attempt to change human nature.

As an example, if construction materials are poorly stacked and they collapse, causing an injury to a worker nearby, the apparent cause could be poor material handling practices. However, possible contributing factors could include employees not understanding the hazard of their actions, the area of storage being unsuitable for the task, or the materials being inadequately maintained. Thus, the true basic causes could identify the need in this case for:

- Further worker training
- Better planning and layout of the storage area
- New equipment or techniques

A thorough accident or incident investigation may require photographs, drawings, and/or technical expertise before the final causes of an accident or incident can be determined and satisfactory controls are chosen.

Basic Steps of an Investigation

When initiating the investigation:

- If workers were injured, make sure they are given appropriate medical attention without delay.
- Control the scene. Place barriers, caution tape, or other devices to prevent others from disturbing the scene.
- Start the investigation as quickly as possible. Conduct interviews at the scene if possible. Ensure that the witnesses discuss the accident or incident in relative privacy so they feel comfortable. Begin with the witnesses who can contribute the most information. These are usually the witnesses that were closest to the accident or incident.
- Take notes during each interview. After the interview, repeat the witness's statement to ensure that you have correctly understood. Document this on a witness statement form and ask the witness to sign the form.
- Close each interview on a positive note and thank all witnesses for their assistance.
- Take immediate corrective action where reasonable to help prevent additional hazards.

- Complete the investigation report with recommendations and follow-up actions.
- Ensure follow-up action occurs.

Key Questions

- **Who?** Get the names of everyone involved, near, present, or aware of possible contributing factors.
- **What?** Describe materials and equipment that were involved, check for defects, get a precise description of chemicals involved, and so on.
- **Where?** Describe the exact location; note all relevant facts (lighting, weather, ground conditions).
- **When?** Note the exact time, date, and other factors (shift change, work cycle, break period).
- **How?** Describe the usual sequence of events and the actual sequence of events before, during, and after the accident or incident.
- **Why?** Find all possible direct and indirect causes and how to keep the event from happening again.

Essential Notions

Causes of accidents or incidents are rarely simple when they are investigated closely. Behind every case there are many contributing factors and causes. The key is to identify those that can be most effectively acted upon to prevent re-occurrences. Investigations should emphasize the long-term eradication of injury, loss, or damage. The focus should be on systems deficiencies in preference to human factors.

After identifying causes and factors, suitable improvement actions must be identified and implemented. Several forms can help the investigator with this task (Figures 12-1 and 12-2).

ACCIDENT / INCIDENT INVESTIGATION REPORT

COMPANY: _____   DATE: _____

ADDRESS: _____

JOBSITE NAME: _____

ADDRESS: _____

NAME OF INJURED	DATE OF ACCIDENT/INCIDENT	TIME OF ACCIDENT/INCIDENT
HOME ADDRESS AND PHONE	EMPLOYEE'S USUAL OCCUPATION	OCCUPATION AT TIME OF ACCIDENT/INCIDENT

EMPLOYMENT CATEGORY	LENGTH OF EMPLOYMENT	TIME in OCCUPATION
☐ Regular, full-time ☐ Seasonal ☐ Temporary ☐ Regular, part-time ☐ Nonemployee	☐ Less than 1 mo. ☐ 6 mos. to 5 yrs. ☐ 1-5 mos. ☐ > 5 years	☐ Less than 1 mo. ☐ 6 mos. to 5 yrs. ☐ > 1-5 mos. ☐ > 5 years

NAME OF OTHER INJURED IN SAME ACCIDENT/INCIDENT

NATURE of INJURY and PART of BODY	TIME of INJURY	SEVERITY of INJURY
	A.M. A._____ P.M. B. Time within shift C. Type of Shift	☐ Fatality ☐ Medical Treatment ☐ First Aid ☐ Other, specify _____

TASK and ACTIVITY at TIME of ACCIDENT/INCIDENT	SUPERVISION at TIME of ACCIDENT/INCIDENT
A. General type of task B. Specific Activity C. Employee was working: ☐ Alone ☐ With crew or fellow worker ☐ Other, specify	☐ Directly Supervised ☐ Not Supervised ☐ Indirectly Supervised ☐ Supervision not feasible

LOCATION OF ACCIDENT/INCIDENT	PHASE OF EMPLOYEES WORKDAY AT TIME OF ACCIDENT/INCIDENT	WEATHER CONDITIONS AT TIME OF ACCIDENT/INCIDENT
ON EMPLOYER'S PREMISES? ☐ Yes ☐ No	☐ During rest period ☐ Performing work duties ☐ During meal period ☐ Entering worksite ☐ Working overtime ☐ Leaving worksite ☐ Other, specify_____	

NAME OF WITNESS TO THE ACCIDENT/INCIDENT

DESCRIBE HOW THE ACCIDENT/INCIDENT OCCURRED

FIGURE 12-1 Sample Accident/Incident Investigation Report (cont.)

ACCIDENT SEQUENCE: Describe in reverse order of occurrence events preceding the injury and accident. Starting with the injury and moving backward in time, reconstruct the sequence of events that led to the injury.

A. Injury Event _____

B. Accident Event _____

C. Preceding Event #1 _____

D. Preceding Event #2, #3, etc. _____

CASUAL FACTORS. Events and conditions that contributed to the accident. Be sure and describe in detail if the proper safety equipment was being used and if it was used correctly.

CORRECTIVE ACTIONS. Those that have been, or will be, taken to prevent recurrence.

Investigation Officer _____   Interpreter _____

Company _____   Company _____

Signature   Date   Signature   Date

© Cengage Learning 2012

FIGURE 12-1 (continued)

WITNESS STATEMENT FORM

WITNESS NAME: _____WITNESS EMPLOYER:_____

ADDRESS: _____

PHONE: _____

AGE: _____OCCUPATION: _____

BRIEF DESCRIPTION OF ACCIDENT/INCIDENT

RELATIONSHIP TO INJURED PARTY

Immediately before the accident, what did you see? Did you notice the injured employee doing anything wrong? Did you warn them? Where were you at? How far away? What did you see?

During the accident, what did you see?

Immediately after the accident, what did you see?

Have you spoken with anyone else concerning this incident?

Additional Comments:

_____   _____
Witness   Date   Investigator   Date

_____
Witness   Date

If you run out of room, use the back of the page

FIGURE 12-2 Sample Witness Statement Form

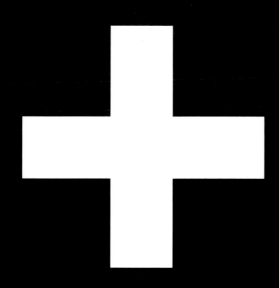

CHAPTER 13

Job Hazard Analysis

In this guide, you have learned many things: the important role of a first aid provider, the concept of REACT, how to care for a patient in cardiac arrest, choking management, the control of bleeding, how to treat shock, treatment for jobsite injuries and illnesses, how to prepare for emergencies, what you need to report to Occupational Health and Safety Administration (OSHA), and how to investigate accidents and incidents. All of that information is extremely important, but employers would clearly prefer that you never had to use that knowledge.

Hazard awareness and hazard prevention can help you achieve your goal of zero injuries. A good starting point for hazard awareness is a job hazard analysis. The following information provided by OSHA relates to the process of a job hazard analysis.

JOB HAZARD ANALYSIS (EXCERPTS FROM OSHA PUBLICATION 3071)

The full version of the OSHA publication can be found at http://www.osha.gov. Questions are directed to the decision makers in your organization.

What Is a Hazard?

A *hazard* is the potential for harm. In practical terms, a hazard often is associated with a condition or activity that, if left uncontrolled, can result in an injury or illness. See Appendix 2 at the back of this book for a list of common hazards and descriptions. Identifying hazards and eliminating or controlling them as early as possible will help prevent injuries and illnesses.

What Is a Job Hazard Analysis?

A *job hazard analysis* is a technique that focuses on job tasks as a way to identify hazards before they occur. It focuses on the relationship between the worker, the task, the tools, and the work environment. Ideally, after you identify uncontrolled hazards, you will take steps to eliminate or reduce them to an acceptable risk level.

Why Is Job Hazard Analysis Important?

Many workers are injured and killed at the workplace every day in the United States. Safety and health can add value to your business, your job, and your life. You can help prevent workplace injuries and illnesses by looking at your workplace operations, establishing proper job procedures, and ensuring that all employees are trained properly. One of the best ways to determine and establish proper work procedures is to conduct a job hazard analysis. A job hazard analysis is one component of the larger commitment of a safety and health management system.

What Is the Value of a Job Hazard Analysis?

Supervisors can use the findings of a job hazard analysis to eliminate and prevent hazards in their workplaces. This is likely to result in fewer worker injuries and illnesses; safer, more effective work methods; reduced workers' compensation costs; and increased worker productivity. The analysis also can be a valuable tool for training new employees in the steps required to perform their jobs safely.

For a job hazard analysis to be effective, management must demonstrate its commitment to safety and health and follow through to correct any uncontrolled hazards identified. Otherwise, management will lose credibility and employees may hesitate to go to management when dangerous conditions threaten them.

What Jobs are Appropriate for a Job Hazard Analysis?

A job hazard analysis can be conducted on many jobs in your workplace. Priority should go to the following types of jobs:

- Jobs with the highest injury or illness rates
- Jobs with the potential to cause severe or disabling injuries or illness, even if there is no history of previous accidents
- Jobs in which one simple human error could lead to a severe accident or injury
- Jobs that are new to your operation or have undergone changes in processes and procedures
- Jobs complex enough to require written instructions

Where Do I Begin?

1. **Involve your employees.** It is very important to involve your employees in the hazard analysis process. They have a unique understanding of the job, and this knowledge is invaluable

for finding hazards. Involving employees will help minimize oversights, ensure a quality analysis, and get workers to "buy in" to the solutions because they will share ownership in their safety and health program.

2. **Review your accident history.** Review with your employees your worksite's history of accidents and occupational illnesses that needed treatment, losses that required repair or replacement, and any "near misses"—events in which an accident or loss did not occur, but could have. These events are indicators that the existing hazard controls (if any) may not be adequate and deserve more scrutiny.

3. **Conduct a preliminary job review.** Discuss with your employees the hazards they know exist in their current work and surroundings. Brainstorm with them for ideas to eliminate or control those hazards. *If any hazards exist that pose an immediate danger to an employee's life or health, take immediate action to protect the worker.* Any problems that can be corrected easily should be corrected as soon as possible. Do not wait to complete your job hazard analysis. This will demonstrate your commitment to safety and health and enable you to focus on the hazards and jobs that need more study because of their complexity. For those hazards determined to present unacceptable risks, evaluate types of hazard controls. More information about hazard controls is found in Appendix 1 located in the back of this book.

4. **List, rank, and set priorities for hazardous jobs.** List jobs with hazards that present unacceptable risks, based on those most likely to occur and with the most severe consequences. These jobs should be your first priority for analysis.

5. **Outline the steps or tasks.** Nearly every job can be broken down into job tasks or steps. When beginning a job hazard analysis, watch the employee perform the job and list each step as the worker takes it. Be sure to record enough information to describe each job action without getting overly detailed. Avoid making the breakdown of steps so detailed that it becomes unnecessarily long or so broad that it does not include basic steps. You may find it valuable to get input from other workers who have performed the same job. Later, review the job steps with the employee to make

sure you have not omitted something. Point out that you are evaluating the job itself, not the employee's job performance. Include the employee in all phases of the analysis—from reviewing the job steps and procedures to discussing uncontrolled hazards and recommended solutions. Sometimes, in conducting a job hazard analysis, it may be helpful to photograph or videotape the worker performing the job. These visual records can be handy references when doing a more detailed analysis of the work.

How Do I Identify Workplace Hazards?

A job hazard analysis is an exercise in detective work. Your goal is to discover the following:

- What can go wrong?
- What are the consequences?
- How could it happen?
- What are other contributing factors?
- How likely is it that the hazard will occur?

To make your job hazard analysis useful, document the answers to these questions in a consistent manner. Describing a hazard in this way helps to ensure that your efforts to eliminate the hazard and implement hazard controls help target the most important contributors to the hazard.

Good hazard scenarios describe:

- Where it is happening (environment)
- Who or what it is happening to (exposure)
- What precipitates the hazard (trigger)
- The outcome that would occur should it happen (consequence)
- Any other contributing factors

A sample form found in Appendix 3 helps you organize your information to provide these details.

Rarely is a hazard a simple case of one singular cause resulting in one singular effect. More frequently, many contributing factors tend to line up in a certain way to create the hazard. Here is an example of a hazard scenario: In the metal shop (environment), while clearing a snag (trigger), a worker's hand (exposure) comes into contact with a rotating pulley. It pulls his hand into the machine and severs his fingers (consequences) quickly.

To perform a job hazard analysis, you would ask:

- **What can go wrong?** The worker's hand could come into contact with a rotating object that "catches" it and pulls it into the machine.
- **What are the consequences?** The worker could receive a severe injury and lose fingers and hands.
- **How could it happen?** The accident could happen as a result of the worker trying to clear a snag during operations or as part of a maintenance activity while the pulley is operating. Obviously, this hazard scenario could not occur if the pulley is not rotating.
- **What are other contributing factors?** This hazard occurs very quickly. It does not give the worker much opportunity to recover or prevent it once his hand comes into contact with the pulley. This is an important factor, because it helps you determine the severity and likelihood of an accident when selecting appropriate hazard controls. Unfortunately, experience has shown that training is not very effective in hazard control when triggering events happen quickly because humans can react only so quickly.
- **How likely is it that the hazard will occur?** This determination requires some judgment. If there have been "near-misses" or actual cases, then the likelihood of a recurrence would be considered high. If the pulley is exposed and easily accessible, that also is a consideration. In the example, the likelihood that the hazard will occur is high because there is no guard preventing contact, and the operation is performed while the machine is running. By following the steps in this example, you can organize your hazard analysis activities.

The following examples show how a job hazard analysis can be used to identify the existing or potential hazards for each basic step involved in grinding iron castings.

Grinding Iron Castings: Job Steps

Step 1—Reach into the metal box to the right of the machine, grasp the casting, and carry it to the wheel.

Step 2—Push the casting against the wheel to grind off any burr.

Step 3—Place the finished casting in the box to the left of the machine.

An example of a Job Analysis Form for Step 1 of the Grinding Iron Cast Scenario above is shown in Figure 13-1. You should repeat similar forms for each job step listed above.

Job Location: *Analyst:*

Metal Shop   Joe Safety   Date:

Task Description: The worker reaches into the metal box to the right of the machine, grasps a 15-pound casting, and carries it to the grinding wheel. The worker grinds 20 to 30 castings per hour.

Hazard Description: Picking up a casting, the employee could drop it onto his foot. The casting's weight and height could seriously injure the worker's foot or toes.

Hazard Controls:

1. Remove the castings from the box and place them on a table next to the grinder.
2. Wear steel-toed shoes with arch protection.
3. Change to protective gloves that allow a better grip.
4. Use a device to pick up the castings.

Task Description: The worker reaches into the metal box to the right of the machine, grasps a 15-pound casting, and carries it to the grinding wheel. The worker grinds 20 to 30 castings per hour.

Hazard Description: Castings have sharp burrs and edges that can cause severe lacerations.

(Continued)

Hazard Controls:

1. Use a device such as a clamp to pick up the castings.
2. Wear cut-resistant gloves that allow a good grip and fit tightly to minimize the chance that they will get caught in the grinding wheel.

Task Description: The worker reaches into the metal box to the right of the machine, grasps a 15-pound casting, and carries it to the grinding wheel. The worker grinds 20 to 30 castings per hour.

Hazard Description: Reaching, twisting, and lifting 15-pound castings from the floor could result in a muscle strain to the lower back.

Hazard Controls:

1. Move the castings from the ground and place them closer to the work zone to minimize lifting. Ideally, place them at waist height or on an adjustable platform or pallet.
2. Train workers not to twist while lifting and reconfigure work stations to minimize twisting during lifts.

FIGURE 13-1 Example Job Hazard Analysis Form

How Do I Correct or Prevent Hazards?

After reviewing your list of hazards with the employee, consider what control methods will eliminate or reduce them. For more information on hazard control measures, see Appendix 1. The most effective controls are engineering controls that physically change a machine or work environment to prevent employee exposure to the hazard. The more reliable a hazard control or the less likely a hazard control can be circumvented, the better. If this is not feasible, administrative controls may be appropriate. This may involve changing how employees do their jobs. Discuss your recommendations with all employees who perform the job and consider their responses carefully. If you plan to introduce new or modified job procedures, be sure employees understand what they are required to do and the reasons for the changes.

What Else Do I Need to Know Before Starting a Job Hazard Analysis?

The job procedures discussed in this booklet are for illustration only and do not necessarily include all the steps, hazards, and protections that apply to your industry. When conducting your own job safety analysis, be sure to consult the OSHA standards for your industry. Compliance with these standards is mandatory, and by incorporating their requirements in your job hazard analysis, you can be sure that your health and safety program meets federal standards. OSHA standards, regulations, and technical information are available online at www.osha.gov.

Twenty-four states and two territories operate their own OSHA-approved safety and health programs and may have standards that differ slightly from federal requirements. Employers in those states should check with the appropriate state agency for more information.

Why Should I Review My Job Hazard Analysis?

Periodically reviewing your job hazard analysis ensures that it remains current and continues to help reduce workplace accidents and injuries. Even if the job has not changed, it is possible that during the review process you will identify hazards that were not identified in the initial analysis.

It is particularly important to review your job hazard analysis if an illness or injury occurs on a specific job. Based on the circumstances, you may determine that you need to change the job procedure to prevent similar incidents in the future. If an employee's failure to follow proper job procedures results in a "close call," discuss the situation with all employees who perform the job and remind them of proper procedures. Anytime you revise a job hazard analysis, it is important to train all employees affected by the changes in the new job methods, procedures, or protective measures adopted.

When Is It Appropriate to Hire a Professional to Conduct a Job Hazard Analysis?

If your employees are involved in many different or complex processes, you need professional help conducting your job hazard analyses. Sources of help include your insurance company, the local fire department, and private consultants with safety and health expertise. In addition, OSHA offers assistance through its regional and area offices and consultation services.

Even when you receive outside help, it is important that you and your employees remain involved in the process of identifying and correcting hazards because you are on the worksite every day and most likely to encounter these hazards. New circumstances and a recombination of existing circumstances may cause old hazards to reappear and new hazards to appear. In addition, you and your employees must be ready and able to implement whatever hazard elimination or control measures a professional consultant recommends.

CHAPTER **14**

First Aid Myths

This chapter could easily be titled, "What Not to Do!" These myths and urban legends regarding first aid treatments are well known, but should not be followed. Let us debunk these first aid myths one last time.

VENOMOUS SNAKE BITE

First Aid Myth

The myth tells us to cut the skin of a snake bite patient in order to suck out the venom. This may be a classic first aid treatment; however, it is thought to be useless at best and may be dangerous. By performing this action, you:

Used under license from Shutterstock.com

- Delay advanced medical treatment
- May cause an infection
- Could cause damage around the wound

Try This Instead

- Follow the steps of "REACT"—call EMS.

    Step 1—**R**ecognize a medical emergency

    Step 2—**E**valuate the environment/Think safety first!

    Step 3—**A**ssess responsiveness

    Step 4—**C**all for EMS

    Step 5—**T**ake action

- Keep the snake bite below heart level.
- Apply a pressure immobilization bandage around the entire length of the bitten extremity if you have been trained.
- Wash the area with warm water and soap.
- Loosen restrictive clothing.
- Suspect and begin treatment for shock while you wait for EMS to arrive.

WHISKEY FOR A TOOTHACHE
First Aid Myth

The myth tells us to drink a shot of whiskey to treat the pain of a toothache. Although many have tried and succeeded in killing pain with alcohol, there are many safer and more effective methods for treating a toothache. By performing this action, you:

- Do not stop the pain; alcohol just stops you from recognizing the pain

Try This Instead

- Rinse your mouth with warm water.
- Use dental floss to remove any food particles.
- Take an over-the-counter (OTC) pain reliever and an OTC antiseptic applied directly to the tooth and gum.
- Contact your dentist if your pain persists for more than a few days, or if you have an infection or fever.

BUTTER ON BURNS
First Aid Myth

The myth tells us to slather a burn with butter, vegetable oil, or other oily substance to help with the pain. It may give you some immediate relief; however, the negative effects can occur quickly. There are several problems with applying butter to a burn:

- Burns damage the skin and butter is not sterile, increasing the chances for infection.
- Butter can actually trap in the heat and cause further damage.
- Butter and other oily substances will need to be cleaned off before the burn can be treated. We all know that oil and water do not mix, causing a very difficult and painful cleaning at the hospital.

Try This Instead—For Minor Burns

- Cool the burn by holding the burn under cool, but not cold, water.
- Cover the burn with a sterile gauze bandage.
- Take an OTC pain reliever as instructed.

JELLYFISH STING

First Aid Myth

The myth tells us to pee on a jellyfish sting. There are no studies that prove that this helps relieve pain from the jellyfish sting. The following problem may occur with this action:

- It is not proven to help the patient and could delay actual proven methods.

Try This Instead

- Follow the steps of "REACT"—call EMS.

  Step 1—**R**ecognize a medical emergency

  Step 2—**E**valuate the environment/Think safety first!

  Step 3—**A**ssess responsiveness

  Step 4—**C**all for EMS

  Step 5—**T**ake action

- Soak in vinegar for about 15 minutes.
- Do not use fresh water or rub the area.
- Remove the tentacles with tweezers.
- Monitor the patient's airway and breathing and be prepared to perform CPR.

NOSEBLEED

First Aid Myth

The myth tells us to put our head back when we get a nosebleed. This can cause more complications than the original nosebleed. The following problems may occur with this action:

- Blood may drain to the throat, making it difficult to breath.
- Blood may drain into the stomach, causing the patient to vomit.

LionelB, 2012. Used under license from Shutterstock.com

Try This Instead

- Have the patient pinch the nostrils together just below the bridge of the nose for about 5 or 10 minutes.
- If available, place an ice pack on the bridge of the patient's nose.
- Do not allow the patient to lie down.
- Call EMS if:
  - There is a foreign object in the nose.
  - Bleeding will not stop.

BEE STING

First Aid Myth

The myth tells us that we need to squeeze the stinger out after being stung by a bee. It is important to remove the stinger, but squeezing can further complicate the situation. The following problems may occur with this action:

- Squeezing the stinger may release venom into the patient's system.
- Allergic reactions may develop.

Try This Instead

- Scrape off the stinger with a credit card or your fingernail as quickly as possible.
- If the patient is allergic to bee stings, call EMS. Assist the patient in taking prescribed medications such as an epinephrine auto-injector.
- Monitor the patient's airway and breathing and be prepared to perform CPR.

SEIZURES

First Aid Myth

The myth informs us that seizing patients will swallow their tongue while seizing. The myth goes on to tell us that putting something in the patients' mouth will prevent them from swallowing their tongue. The following problems may occur with this action:

- Putting something in their mouth can actually block the airway.
- You may lose your finger if the seizing patient clamps down on it.

Try This Instead

- Follow the steps of "REACT"—call EMS.

  Step 1—**R**ecognize a medical emergency

  Step 2—**E**valuate the environment/Think safety first!

  Step 3—**A**ssess responsiveness

  Step 4—**C**all for EMS

  Step 5—**T**ake action

- Keep the patient calm, and provide comfort and care.
- Move all objects that could hurt the patient out of the way.

BLACK EYE

First Aid Myth

The myth tells us to slap a raw steak on your black eye. The following problems may occur with this action:

- Grease or other contaminants (*E. coli*) from the steak may get into your eye, causing an infection.
- You just ruined a perfectly good steak.

Try This Instead

- Apply a cold compress or bag of frozen peas to the eye.

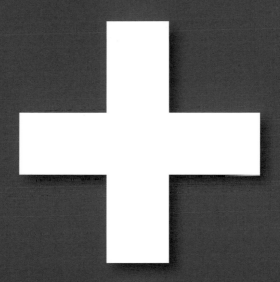

Hazard Control Measures

Information obtained from a job hazard analysis is useless unless hazard control measures recommended in the analysis are incorporated into the tasks. Managers should recognize that not all hazard controls are equal. Some are more effective than others at reducing the risk.

The following list represents the order of precedence and effectiveness of hazard control:

1. Engineering controls
2. Administrative controls
3. Personal protective equipment

Engineering controls include the following:

- Elimination/minimization of the hazard; designing the facility, equipment, or process to remove the hazard; or substituting processes, equipment, materials, or other factors to lessen the hazard
- Enclosure of the hazard using enclosed cabs, enclosures for noisy equipment, or other means
- Isolation of the hazard with interlocks, machine guards, blast shields, welding curtains, or other means
- Removal or redirection of the hazard such as with local and exhaust ventilation

Administrative controls include the following:

- Written operating procedures, work permits, and safe work practices
- Exposure time limitations (used most commonly to control temperature extremes and ergonomic hazards)
- Monitoring the use of highly hazardous materials
- Alarms, signs, and warnings
- Buddy system
- Training

Personal protective equipment—such as respirators, hearing protection, protective clothing, safety glasses, and hard hats—is acceptable as a control method in the following circumstances:

- When engineering controls are not feasible or do not totally eliminate the hazard
- While engineering controls are being developed
- When safe work practices do not provide sufficient additional protection
- During emergencies when engineering controls may not be feasible

Use of one hazard control method over another that is higher in the control precedence may be appropriate for providing interim protection until the hazard is permanently abated. In reality, if the hazard cannot be eliminated entirely, the adopted control measures will likely be a combination of all three items instituted simultaneously.

Common Hazards and Descriptions

Hazard	Hazard Description
Chemical	
Toxic	A chemical that exposes a person by absorption through the skin, inhalation, or passage through the bloodstream that causes illness, disease, or death. The amount of chemical exposure is critical in determining hazardous effects. Check the Material Safety Data Sheets (MSDS) and/or OSHA 1910.1000 for chemical hazard information.
Flammable	A chemical that, when exposed to a heat ignition source, results in combustion. Typically, the lower a chemical's flash point and boiling point, the more flammable the chemical. Check the MSDS for flammability information.
Corrosive	A chemical that, when it comes into contact with skin, metal, or other materials, damages the materials. Acids and bases are examples of corrosives.
Electrical	
Shock/Short Circuit	Contact with exposed conductors or a device that is incorrectly or inadvertently grounded, such as when a metal ladder comes into contact with power lines. For example, 60 Hz alternating current (common house current) is very dangerous because it can stop the heart.
Fire	Use of electrical power that results in electrical overheating or arcing to the point of combustion or ignition of flammables, or electrical component damage.

Hazard	Hazard Description
Electrical (*cont.*)	
Static/ESD	The moving or rubbing of wool, nylon, other synthetic fibers, and even flowing liquids can generate static electricity. This creates an excess or deficiency of electrons on the surface of material that discharges (spark) to the ground, resulting in the ignition of flammables or damage to electronics or the body's nervous system.
Loss of Power	Safety-critical equipment failure as a result of loss of power.
Ergonomics	
Strain	Damage of tissue due to overexertion (strains and sprains) or repetitive motion.
Human Error	A system design, procedure, or equipment that is error-provocative. (A switch goes up to turn something off.)
Excavation (Collapse)	Soil collapse in a trench or excavation as a result of improper or inadequate shoring. Soil type is critical in determining the hazard likelihood.
Explosion	
Chemical Reaction	Sudden and violent release of energy caused by the interaction of chemicals with each other or the environment.
Over Pressurization	Sudden and violent release of a large amount of gas/energy due to a significant pressure difference such as a rupture in a boiler or compressed gas cylinder.
Fall (Slip, Trip)	Conditions that result in falls (impacts) from height or traditional walking surfaces (such as slippery floors, poor housekeeping, uneven walking surfaces, exposed ledges, etc.).
Fire/Heat	Temperatures that can cause burns to the skin or damage to other organs. Fires require a heat source, fuel, and oxygen.

Hazard	Hazard Description
Mechanical	Skin, muscle, or body part exposed to crushing or caught-between cutting, tearing, or shearing items or equipment.
Mechanical Failure	The breakdown of equipment; typically occurs when devices exceed designed capacity or are inadequately maintained.
Mechanical/Vibration (Chaffing/Fatigue)	Vibration that can cause damage to nerve endings, or material fatigue that results in a safety-critical failure. (Examples are abraded slings and ropes, weakened hoses, and belts.)
Noise	Noise levels (>85 dBA 8 hr TWA) that result in hearing damage or inability to communicate safety-critical information.
Radiation	
Ionizing	Alpha, Beta, Gamma, neutral particles, and X-rays that cause injury (tissue damage) by ionization of cellular components.
Non-Ionizing	Ultraviolet, visible light, infrared, and microwaves that cause injury to tissue by thermal or photochemical means.
Struck Against	Injury to a body part as a result of coming into contact with a surface in which action was initiated by the person. (An example is when a screwdriver slips.)
Struck By (Mass Acceleration)	Accelerated mass that strikes the body causing injury or death. (Examples are falling objects and projectiles.)
Temperature Extreme (Heat/Cold)	Temperatures that result in heat stress, exhaustion, or metabolic slowdown such as hypothermia.
Visibility	Lack of lighting or obstructed vision that results in an error or other hazard.
Weather Phenomena (Snow/Rain/Wind/Ice)	This is self-explanatory.

Sample Job Hazard Analysis Form

Job Title:	Job Location:	Analyst	Date
Task #	Task Description:		
Hazard Type:	Hazard Description:		
Consequence:	Hazard Controls:		
Rationale or Comment:			

NURSING ESSENTIALS
Drugs